HORSE
NUTRITION

PRENTICE
HALL PRESS
EQUESTRIAN
BOOKS

HORSE NUTRITION

• A PRACTICAL GUIDE •

Harold F. Hintz, Ph.D.

PRENTICE
HALL
PRESS

New York   London   Toronto   Sydney   Tokyo   Singapore

Prentice Hall Press
15 Columbus Circle
New York, New York 10023

Published in 1988 by the Prentice Hall Trade Division

PRENTICE HALL PRESS and colophon are registered
trademarks of Simon & Schuster Inc.

Originally published by Arco Publishing, Inc.

Library of Congress Cataloging-in-Publication Data

Hintz, Harold Franklin, 1937–
   Horse nutrition:

   Includes bibliographies.
   1. Horses—Feeding and feeds.   I. Title.
SF285.5.H56 1983   636.1'084   82–16294
ISBN 0–668–05416–6
Manufactured in the United States of America

10 9 8 7 6 5 4

*This book is dedicated to
Red, Pal, E.J., and Freckles—
four of the ponies who helped us get
the Equine Research Program at Cornell underway.*

Contents

Acknowledgments   ix
Introduction   xi

Chapter One.   A Brief History of the Feeding of Horses   1
                    Grains   1
                    Forages   3
                    Protein Supplements   6
                    Commercial Tonics and Feeds   7

Chapter Two.   The Digestive System   12
                    The Digestive Tract   12
                    Digestibility of Feeds   18

Chapter Three. Nutrients   31
                    Energy   31
                    Energy Evaluation   32
                    Protein   35
                    Minerals   39
                    Toxic Minerals   63
                    Vitamins   66

Chapter Four.   Feeds   84
                    Forages   84
                    Silage   111
                    Straws   112
                    Pasture   113

Grains   121
Other High Energy Sources   129
Root Crops   131
Other Feeds   132
Miscellaneous Feeds   136
Protein Supplements   137
Commercial Feeds   140
Feed Preferences   147

Chapter Five.   Feeding Programs   154
                How Much Hay Does a Horse Need?   155
                The Mature Horse   158
                The Young Horse   178
                Growth Rates for Young Horses   188
                Performance Horses   194
                Donkeys and Mules   204

Chapter Six.   Metabolic Problems and Other Concerns   207
                Founder   207
                Exertional Myopathy   210
                Hyperlipemia   212
                Eclampsia   213
                Epistaxis   213
                Heaves   215
                Hoof Growth and Nutrition   215
                Sweet Feed Bumps   218
                Hair Coat   219
                Temperament and Behavior   220
                Anemia   220

General Bibliography   223

Index   224

Acknowledgments

I would like to thank Herbert Schryver, Jack Lowe, Clarence Marquis, Vince Soderholm, Janice Williams, Pete Daniluk, Eileen Burgess, and my family for all their help.

Acknowledgements

Introduction

Every year the market is flooded with information on horsefeeding—in books, magazines, bulletins, manufacturers' pamphlets, and advertisements. Why, then, am I adding to this list? Every horse lover knows that a horse must be fed properly to achieve its genetic potential, but the available material on the subject is widely scattered. It is difficult to find a single reliable source. Some of the information may be biased or based on limited interpretations and observations. There are always honest differences of opinion, and more information is being supplied continuously as the result of various experiments.

I hope that my twenty years of experience with the Equine Research Program at Cornell University will help you cut through this array of information and consolidate your own observations. I deal with digestion, nutritional requirements, feeds, feeding programs, and nutrition-related problems to help you evaluate feeding practices in a logical, practical, and economical manner. I also discuss some of the lingering myths and misconceptions that provide colorful anecdotes but can harm horses.

This book is primarily for the horse owner, but I hope it will also interest and benefit the manufacturers of horse feeds, veterinarians, teachers and students of horse feeding, and all those interested in comparative nutrition.

It is a tribute to horses that they have managed to do so well in spite of humans. We have taken them from a free-ranging environment, placed them in small stalls, forced them to perform intensively while still youngsters, changed their breeding time according to an artificial calendar, and then we wonder why our horses have problems.

Clearly this book will not contain all of the answers; many of them I don't know and many of them nobody knows, and there is always a point at which the "eye" of the horse owner takes over. Dairy cattle research has been conducted at almost every experiment station for more than 60 years, yet we still have much to learn. Horse research was discontinued at most experiment stations when the combustion engine replaced the draft animal. Research was reintroduced over a decade ago at a handful of agricultural experiment stations, but much more research is needed before we know all the answers.

The first chapter of this book deals with the history of horse feeding; this provides an interesting perspective and will help you understand the evolution of feeding practices.

Chapter Two examines the digestive tract, so very different in equines from all other farm animals. It is essential for you to know how the digestive system works to understand why feeds have different values and how feeds are utilized. The incidence of many problems such as colic, founder, impactions, and diarrhea can be avoided through the development of improved feeding practices and management.

Nutrients are covered in Chapter Three. Functions, toxicity levels, deficiency signs, sources of nutrients, and methods of diagnosing and preventing nutritional problems are all examined closely.

Characteristics of some of the great variety of feeds are outlined in Chapter Four.

Chapter Five discusses the feeding programs for the various classes of horses, mules, and donkeys developed at experiment stations and by horse owners. Advice is given on how to formulate your own feeds and evaluate commerical feeds.

Metabolic and nutrition-related problems such as founder, colic, azoturia, eclampsia, hyperlipemia, feed bumps, epiphysitis, and contracted tendons are explored in Chapter Six. Nutritional factors affecting hoof growth and hair coat are also analyzed.

Several references are included at the end of each chapter for those who desire more detailed information. The books cited in several different chapters are also listed in the general bibliography for your interest and further reading.

Chapter One

A Brief History of the Feeding of Horses

The relationship between humans and horses probably started with humans eating horses rather than feeding them. For example, the bones of thousands of horses—estimated to be at least 20,000 years old—were found outside a rock shelter near Solutre, France. It is believed that the animals from which the bones came were hunted for food.

It is difficult to determine when the horse was first domesticated, but historians have suggested that domestication probably took place in west-central Asia about 3000 B.C. The first "horseman" probably did little in the way of feeding beyond allowing his horses access to forage.

Grains

Barley is usually considered to be the first crop cultivated by man, so it was probably the first grain fed to horses. Barley was probably domesticated in southwestern Asia between 9000 B.C. and 7000 B.C. Wheat was also available by the time the horse was domesticated, but it was primarily used for human consumption. Oats were domesticated much later than barley or wheat, perhaps not until 1000 B.C. Corn, a New World crop, would not have been available to ancient horsemen.

Barley appears to be the grain most commonly fed to horses in antiquity. We know from the Bible (1 Kings 4:28) that in King Solomon's time—about 1000 B.C.—horses were fed barley. And it was recorded that the emperor Caligula fed his horse barley from a golden cup in A.D. 40. Columella, in A.D. 50, recommended the feeding of barley but suggested that a mixture of roasted wheat and wine would fatten a thin horse quicker than barley.

By 3000 B.C. barley had reached the British Isles and was soon the predominant grain. It was not until A.D. 700, however, that oats were grown in quantity in England. Oats ultimately became the favorite of English horse owners. By 1840 John Stewart could write, "In this country [England] horses are fed upon oats, hay, grass and roots. People talk as if they could be fed on nothing else." Although some barley was fed in nineteenth-century England, it was not considered as economical as oats.

Corn was introduced from North America but was not widely accepted. According to one account in 1880: "There is a prejudice against it [corn] which has prevented it being tried long enough to enable us to form a good opinion of its merits. Bracy Clarke claims it clogs the stomach and tends to produce founder. But perhaps he forgot that any nutritious grain can produce founder." Other nineteenth-century writers reported that "corn tends to make animals fat and liable to sweat."

When Samuel Johnson defined *oats* in his dictionary, he said they were eaten by people in Scotland but only by horses in England. To which a Scot might reply, "That's why England has such good horses and Scotland such fine men."

European settlers in North America brought with them wheat, rye, barley, and oats to cultivate. At first these crops were grown more extensively than corn. Oats were grown for horses (except that Scottish settlers grew oats for their own consumption), barley for beer, and wheat and rye for bread.

Corn became the major grain raised in North America but, as in England, many horse owners did not like it. It was said that "Corn causes horses to become fat and does not give the hard muscular flesh which oats do, hence their softness and want of endurance." But corn found favor with others. In a letter to *The American Agriculturist* in 1866 a farmer wrote that "Many object to the heating properties of corn, yet experiments prove corn to be the best daily for a hard-working horse; corn may be heating simply because it is very nutritious."

In 1898, W. A. Henry in *Feeds and Feeding* (which, in later editions, F. B. Morrison co-authored and finally authored) quoted a German scientist who said, "Corn tends to make the animal fat and liable to sweat. While it improves their appearance, it somewhat detracts from their physical energy." But Henry reviewed experiments in which corn-fed horses did quite well; he conceded that "Corn is not the equal of oats as a grain for the horse; nevertheless, because of its low cost and the high feeding value it possesses, [it] will be extensively used where large numbers of horses must be economically maintained."

In 1932 Dr. Carl Gay wrote:

> When its general use in the Corn Belt states is considered, much of the prejudice of the Eastern feeders [against corn] loses weight. The average Iowa horse, for instance, is produced by a dam which was raised on corn, and had no other grain during the period of carrying and suckling her foal. The foal receives a little cracked corn or even cob corn for his first bite, with the amount gradually increased until he is allowed from 20 to 40 ears per day at maturity. In spite of this fact, when these very horses come East, top our markets, and pass under the management of the city stable boss, corn is absolutely prohibited as dangerous to feed; yet it requires a long time to induce and teach some of these horses to eat anything else.

Nevertheless, oats continued to be the grain horsemen preferred to corn because they are highly palatable to horses, have a higher fiber content, are a much safer feed, and contain a higher level of protein.

Forages

Alfalfa is thought to be the first forage cultivated by man and therefore the first cultivated forage fed to horses by man. Alfalfa was carried into Greece by the Persians during Xerxes' invasion in 490 B.C. Alfalfa was planted to raise feed for the war horses. The

after conquering Greece, brought alfalfa to Rome in 146

know that Columbus carried alfalfa to the West Indies and that Spanish explorers took alfalfa to South America, where it was soon grown extensively. Although cultivated somewhat in the colonial times, alfalfa was not a popular crop. Both George Washington and Thomas Jefferson tried unsuccessfully to raise alfalfa. The lack of knowledge of the pH soil requirement was perhaps the problem. In *Feeds and Feeding* Henry wrote, "Attempts to grow alfalfa in the Eastern States have generally ended in failure." And as recently as 1906, one writer said that alfalfa was a "new" crop to New England.

Alfalfa was brought to southern California by the Spanish, but the real impetus for alfalfa production in the West was the seed brought by 49'ers who had seen the productive fields in South America as they traveled the Cape Horn route from the East Coast to the gold fields. Alfalfa production quickly spread from California to other western states. Thus substantial amounts of alfalfa were produced in the West at a time when it was little known in the East. In 1900 only 1 percent of all U.S. alfalfa was grown east of the Mississippi River.

The value of alfalfa for horses has been debated for years. Although the ancients were unanimous in their support of alfalfa and many writers in the nineteenth and twentieth centuries praised it, many others were—and are—opposed to its use. Among the negative comments are that alfalfa causes horses to become fat, sweat excessively, lose endurance and strength, and to have kidney problems. However, Rham's *Dictionary for the Farm* (1853) commented: "An acre of lucerne (alfalfa) will keep four or five horses from May to October. Horses can work hard upon it without any grain provided it be slow work."

Joseph E. Wing (Alfalfa Joe), a strong advocate of alfalfa, brought alfalfa seed from Utah to his farm in Ohio. In 1912 he reported that he fed all of his horses alfalfa and that they had no trouble with colic or kidney damage. "Idleness and excessive alfalfa will make a horse soft," he said. "Idleness and six eggs a day will make all sorts of things wrong with a man, for that matter. A hard horse can be developed on alfalfa if the horse is trained properly.

I feed alfalfa because the horses like it better than timothy and because I can save grain." Wing also believed that "Alfalfa is often fed in too large amounts to horses and the excess of nutriment fed then must be eliminated and that fact makes them sweat more and tire sooner than had they not been overfed. The plain truth is that timothy is safest for horses because it is not much more than a filler."

But in spite of the many proponents of alfalfa, timothy hay remained the favorite hay of many horsemen, particularly those in the eastern United States. Timothy is a safer feed than alfalfa because horses are much less likely to overfeed on it. Alfalfa is much more leafy than timothy, and in the early days, it was more difficult to make hay from alfalfa that was free from dust and mold than it was from timothy. Furthermore, mature horses really didn't need the high level of protein and energy contained in alfalfa, so it was much more logical for the farmer to feed the alfalfa to his cattle and the timothy hay to the horses. According to Henry, "One reason for the popularity of timothy lies in the fact that it is easily distinguished from hay of all other grasses and both the farmer who grows it and the horseman who feeds it feel no uncertainty as to its identity or quality. Its freedom from dust commends good timothy hay as a horse feed and it is an excellent roughage for horses, whose sustenance comes mostly from grain."

In the 1920s and 1930s many experiments were conducted with draft horses in which corn was compared to oats or alfalfa hay was compared to timothy hay.

In 1925 R. S. Hudson of the Michigan Agricultural Experiment Station compared alfalfa and timothy for draft horses. Nine teams were used in the year-long study. It was discovered that the horses fed alfalfa worked an average of 236 of the possible 300 work days in the year. Those fed timothy worked an average of 232 days. The average weight of the horses fed alfalfa was 1,692 pounds; they ate 12.2 pounds of corn and 17.9 pounds of alfalfa hay daily. Those fed timothy weighed 1,696 pounds and ate 8 pounds of corn, 6.2 pounds of oats, and 19.6 pounds of timothy per day. It cost 31 cents per day for feed, or 7 cents for an hour of work for the timothy-fed horses. No problems could be attrib-

Table 1.1.   Feed and Bedding Formula Data Taken
in Wayne and Monroe Counties on 23 Farms
Representing 118 Horses for Year Ending
May 1, 1921

Average Direct Costs for a Horse for 1 Year	
Expenses	
Oats—1,773 pounds at $20.00 per ton	$17.73
Corn—860 pounds at $15.00 per ton	6.45
Corn Stover—631 pounds at $5.00 per ton	1.58
Hay Mixed—10,113 pounds at $12.00 per ton	60.68
Pasture—3.5 weeks at $0.40 per week	1.40
Bedding—1,377 pounds at $4.00 per ton	2.75
Total Feed and Bedding Costs	$90.59

uted to the diets. It was concluded that "Alfalfa is a safe, efficient, and economical feed for work horses."

The modern horseman might be interested in other early reports about the cost of keeping horses and the rations fed. Table 1.1 shows a summary of a survey conducted in Michigan in 1921. The cost of keeping horses certainly has increased since then.

Many of the farmers who moved West carried their preference for timothy with them and were often willing to pay a higher price for grass hay than for alfalfa.

Red clover hay was discriminated against by horse owners of the early 1900s because it was usually loaded with dust and often cut when too mature. But many horse owners in the 1940s and 1950s began to appreciate the value of good-quality red clover hay and it was frequently in high demand.

Protein Supplements

Alfalfa hay was the earliest high protein source, fed to the horses of antiquity. Although the ancients did not know about protein or its importance, they realized that certain feeds gave certain responses. Columella recommended the use of horsebeans (*Vicia faba*) in A.D. 50. Beans continued to be used thereafter, and many English horsemen of the 1700s and 1800s recommended the use of steamed or cooked horsebeans and other beans. Whole flaxseed was used; it not only provided protein but fat as well.

The development of the vegetable oil industry produced protein-rich by-products from the extraction of the oil. Cottonseed meal and linseed meal were used extensively for horses in the late 1800s and early 1900s. Linseed meal was particularly prized because it was thought to impart "bloom and finish to the hair coat," and because it was a laxative. The importance of protein was demonstrated in trials at the University of Maine in 1891 which clearly showed that the addition of peas or linseed meal to a grass hay-oat ration for young horses increased the rate of gain of Percheron colts. Other studies in the 1890s showed that the addition of peanut meal and coconut meal could also improve growth rate. A diet of grass hay and oats would not contain adequate levels of protein for rapid growth.

Soybeans appear in Chinese writings as early as 2838 B.C. and were probably raised long before that. But in the early 1900s only limited acres of soybeans were grown in the United States. At that time soybeans were one of the most important agricultural crops of northern China and Japan, and a considerable amount of soybean meal was imported to the Pacific Coast states from the Orient. The imported meal was used primarily for dairy cattle and poultry rations. Many horsemen continued to prefer linseed meal to soybean meal, but this was subsequently reversed; soybean meal is more popular today.

Commercial Tonics and Feeds

Not many horse owners in the United States fed commercially mixed or prepared feeds prior to 1900. Most farmers fed hay and grain raised on the farm. Nonfarming horse owners simply bought hay and grain from the feed store. In 1895 less than 2 percent of the brands of feed found on sale in New England were mixed feeds.

But many tonics and other concoctions were available. Thorley's Feed in 1860 contained horse beans, barley, flaxseed, Peruvian bark, and quinine tonic, and cost 10 times as much per pound as oats. Raven's Horse, Cattle and Poultry Food of 1893 was claimed to be the best tonic, blood purifier, and system regulator known. It was claimed to make hens lay, cure cholera and roup, and to be excellent for breeding animals; it cost about twenty times the price of oats. Another tonic claimed to "cure all diseases of horses, besides acting as a tonic to the sexual organs of stallions; prevents bloat, flatulence, and cures diarrhea and constipation."

Many of the tonic feeds of the late 1800s and early 1900s contained herbs and spices such as gentian, fennel, fenugreek, anise, asafetida, cayenne, cinnamon, ginger, sulfur, mustard seed, turmeric, and cocoa shells. When additions were made to the tonics, they were often given exotic names—corn meal became "fermention powder," black pepper was "capsicum," and dried paint was "princess metallic."

But the medicine-show claims made for the tonics were disputed by the faculties of agricultural colleges and various experiment stations.

A Louisiana Experiment Station Bulletin of 1906 summed them up: "These tonics are too expensive. If your animals are in good health they need no tonic and if they are sick it is cheaper to consult a veterinarian." A 1905 bulletin from New Jersey said "The claims of the manufacturers of condimental feeds when not preposterous, are exaggerated and misleading." In spite of the warnings, tonics were still commonly used until the 1930s, when many states made the sale of such feeds illegal.

Similar control programs were introduced to curb the large number of other unreliable products and fraudulent practices. For example, one company sold ground wheat that was adulterated with finely ground corn cobs. Low-quality grain, i.e., damaged corn, shriveled wheat, and low-quality by-products such as wheat screenings and oat hulls were frequently found in mixed feeds in the early 1900s. Heilman's mixed feed of 1898 was found to contain a considerable amount of sawdust. But even Heilman's was outdone. Another company sold material containing ground-up brush, bark, and twigs to which twenty gallons of molasses was added per ton. The H.O. Horse Feed of 1902 was primarily oats, corn, wheat, and linseed meal; Wilbur's Seed Meal Horse Food, on the other hand, was composed primarily of charcoal, gluten meal, and linseed meal and cost more than twice as much as H.O. Horse Feed.

In 1899 Connecticut, Maine, and Vermont were the only states that had laws requiring the seller to guarantee protein and fat contents of their products. By 1923 all states except Montana had feed inspection laws requiring "the manufacturer to state what he sells and sell what he states." In the first years of inspection in some states more than one half of mixed feed samples were not "what was stated," and the production of some poor-quality feeds persisted. The Kentucky Bulletin of 1923 reported:

> Many of the feeds in this group [horse feeds] are made from the best materials obtainable and show care and intelligence in their manufacture, but, on the other hand, many are mixtures of low-grade materials and are often sold at excessive prices. Some of the brands are sold under fancy and catchy names and under names that are in no sense descriptive of the feeds or materials used. Molasses is added in many cases to cover up inferior material.

Incidentally, significant use of molasses in commercial horse feeds started almost with the production of horse feeds.

Fortunately, the quality of commercial horse feeds improved with time. The New York State Agricultural Station in 1927 reported, "A large proportion of the commercial horse feeds sold on the market are composed of cracked corn, whole, ground or crushed oats, ground alfalfa, and molasses. There are a few horse feeds composed of materials such as wheat bran, oat hulls, and hominy feed." In 1933 the Kentucky Department of Agriculture reported that the horse and mule feeds registered in that state usually consisted of corn, oats, molasses, and wheat bran.

By-products such as yeasts containing high levels of B vitamins were added to commercial rations in the 1920s and vitamin-rich mixtures were added in the 1930s.

Calcium was first added routinely to commercial horse feeds in the mid-1930s. The delay is difficult to understand because it had been known for years that a lack of calcium in the diet causes skeletal diseases such as nutritional secondary hyperparathyroidism (NSH). As early as A.D. 400 Vegetius described a condition he called animal osteomalacia, which was probably NSH. But it wasn't until the end of the nineteenth century that calcium deficiency was *identified* as the cause of NSH.

Horses with NSH frequently develop "big head" because fibrous connective tissue invades the head. "Big head" must have been a common problem in the 1800s, judging from the large number of letters and articles in journals and magazines. Many of the early treatments were quite severe. Even in the late nineteenth century veterinarians were recommending that the "growth" be cut out and four grains of arsenic be put in the wound. Removal of teeth was also frequently recommended.

Sometimes the treatments were effective even though the theory behind the treatment was incorrect. In 1855 Dr. A. S. Copeman claimed that big head was a fatty degeneration caused by excessive use of "hydrocarbonaceous" feeds such as grains. The horses improved when he removed grain from the diet and fed them only hay. Of course, grain contains almost no calcium. Hays, particularly those with some legume, can supply a good deal of calcium.

In the 1960s commercial feeds took a new form as pelleted grains and pelleted complete feeds started to make an impact. European war horses had been fed biscuits in the 1870s and the German armies used compressed feeds in World War I and II. However, it was not until the 1950s that research on pelleted feeds for horses was started in the United States. The large increase in numbers of suburban horse owners created a demand for the new form of horse feed, and the sales of pelleted feeds increased rapidly.

References

Allen, A.B. *Notes to the Stable Book of J. Stewart.* New York: Orange Judd Co., 1885.

Columella. *De re rustica.* Translated by H.B. Ash, London, 1941–55.

Copeman, A.S. Big-head disease. *Amer. Vet. J.*, 1:217, 1855.

Hudson, R.S. Alfalfa and horses. *Mich. Agr. Exp. Stat. Quart. Bull.,* 7:75, 1925.

Kellner, O. *The Scientific Feeding of Animals.* Translated by W. Goodwin. New York: Macmillan Co., 1910.

Stewart, J. *The Stable Book.* Edinburg: Blackwood and Sons, 1841.

Walsh, J.H. *The Horse in the Stable and the Field.* London: Routledge, Warne, and Routledge, 1880.

Wing. J.E. *Alfalfa in America.* Chicago: Sanders Publ. Co., 1912.

Young, A. Review of new publications relating to agriculture. *Annals Agric.,* 1:123, 1780.

———. A fortnight's tour in Kent and Essex. *Annals Agric.,* 2:42, 1783.

Chapter Two

The Digestive System

The Digestive Tract

The digestive tract of the equine is different from that of any other domestic animal. The digestive tracts of the rabbit and guinea pig are similar to the horse but even they have significant differences. Knowledge of the anatomy of the digestive tract and understanding of the digestive physiology of the horse is essential for sound health and feeding practices. The gastrointestinal tract is illustrated on page 13.

MOUTH

Proper utilization of feed requires a sound mouth, or management practices to overcome an unsound mouth. Routine dental care is an important part of management. Because of the jaw movement of the horse the back teeth may develop sharp projections that interfere with chewing or may even injure the tongue and cheeks. A horse with dental problems will often slobber excessively and lose feed. The projections can be filed down (floated) with a rasp. Some horses may never need to have their teeth floated, whereas others may need it every six months; thus, routine inspection is advisable. Some extra teeth or caps may need to be

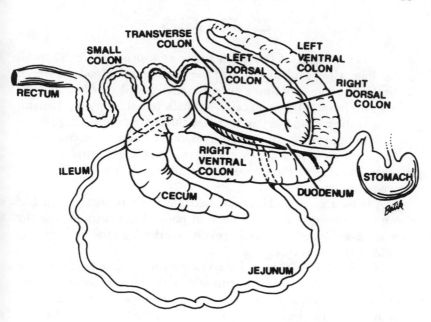

Figure A. Diagram of the digestive tract of the horse. (Courtesy W. F. Schryver and the author)

removed. If the horse's teeth problems cannot be corrected by an equine dentist or veterinarian, the animal should be given appropriate feeds. Crushed or rolled grains should be fed rather than whole grains. Hay can be chopped. Complete pelleted feeds can be used. If the pellets are too hard, perhaps they can be softened with water or even made into a gruel.

Sometimes more drastic measures are needed. A few years ago a 31-year-old donkey in a Japanese zoo was having difficulty eating because many of his teeth were missing. As the donkey was such a favorite of the children, the zoo spent $2,200 to have dentures attached to his remaining incisors. The donkey returned to good health. Obviously most owners cannot afford dentures, but they can afford routine dental care.

ESOPHAGUS

The esophagus of the adult horse is approximately 50 to 60 inches long. Choke (obstruction of the esophagus by food or foreign bodies) is the most common nutritionally related problem and occurs most frequently in greedy animals fed dry grains. Inflammation of the esophagus can increase the chances of choke; ponies may be more prone to this than horses. Some horses have congenital or acquired stenosis (narrowing of the esophagus) that requires surgery. Decreasing the rate of intake and moistening the feed can alleviate the problem.

A horse that is choking may become anxious, may get up and down continuously, arch his neck in painful spasms and paw the ground. Food and saliva are regurgitated through the nostrils, and coughing can be severe.

A veterinarian should be consulted immediately if choke is suspected, but don't panic. Fortunately choke in the horse does not cause asphyxiation as it does in humans, because in humans choke usually refers to blockage of the trachea rather than blockage of the esophagus. If, however, the wad remains in the esophagus for several hours it may cause pressure necrosis (death of the cells of the esophagus). Sometimes no treatment is needed, since the wad of food may be softened by saliva and eventually slip down the esophagus; sometimes tranquilizers can be given to relax the horse and release the wad. Do not allow the horse to have access to feed or water while waiting for the veterinarian to come. The veterinarian can sometimes gently push the obstruction into the stomach with a stomach tube. If not, more drastic measures such as surgery may be necessary.

STOMACH

The stomach of the horse provides approximately 8 percent of the volume of the gastrointestinal tract. The stomach mixes the food, initiates digestion with secretions such as hydrochloric acid, and serves as a reservoir to help maintain a constant supply of material to the small intestine. Food moves fairly rapidly through the stomach of the horse compared to some other species. Because the stomach is small and because the horse seldom if ever vomits, the horse should be fed three or more times daily if large amounts of grains are involved. Severe overfeeding can cause colic or even rupture of the stomach.

Gastric ulcers can be a problem, particularly in young horses, but the relationship of nutrition to ulcers is not clear. Stress is probably of greater significance than diet as a cause of gastric ulcers, but diet interaction cannot be ruled out.

In 1982 Dr. W. Rebhun and co-workers diagnosed gastric ulcers in five foals. The signs were depression, colic, odontroprisis (grinding of teeth), excessive salivation, and a strong tendency to lie on the back—apparently in an effort to relieve pain. The foals were treated with antacids or cimetidine (a histamine antagonist used in the treatment of gastric ulcers in humans) and were fed rations high in fiber and low in grain. Three of the five foals died and numerous ulcers were found at necropsy. One of the foals showed marked improvement and apparently recovered without any after-effects. The other surviving foal improved gradually and showed no signs of gastric ulceration, but remained small for her age.

SMALL INTESTINE

The small intestine provides about 30 percent of the capacity of the gastrointestinal tract. The first segment of the small intestine is called the duodenum, the lengthy middle section is called the jejunum, and the last part the ileum (see Figure A). Enzymes that digest carbohydrate, protein, and fat are secreted into the lumen of the small intestine. The small intestine is the primary site of soluble carbohydrate, fat, and protein digestion. Soluble carbohydrates such as starch are hydrolized into glucose and other simple sugars which can be absorbed and utilized for energy. Proteins are hydrolized into amino acids; amino acids are absorbed and used as building blocks for the body.

Fat digestion in the small intestine is aided by bile secretion. It has been suggested (erroneously) that horses cannot digest fat

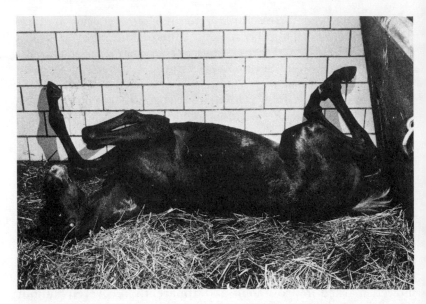

Foals with gastric ulcers have a strong tendency to lie on the back. (Courtesy of William Rebhun, Cornell University)

because the horse does not have a gallbladder. The gallbladder in species such as man stores bile and releases it in response to hormones triggered by the presence of fat in the intestine. But, although he has no bladder, the horse can digest fat efficiently and apparently secretes a continuous flow of bile. Incidentally, many other animals such as the rat, deer, elk, moose, rhinoceros, giraffe, camel, llama, tapir, and elephant do quite nicely without a gallbladder.

The small intestine is also the primary site of absorption of many minerals such as calcium, zinc, copper, magnesium, and manganese. Phosphorus and the electrolytes can be absorbed from the small or large intestine.

The small intestine can become impacted with parasites (ascarids), digesta, or foreign objects. However, impaction of the small intestine is not as common as impaction in the large intestine. This is because the small intestine is of relatively uniform size, whereas the large intestine has several narrows and bends (see Figure A); also, the contents of the small intestine are usually more liquid than those of the large intestine.

LARGE INTESTINE

The large intestine consists of the cecum, right ventral colon, left dorsal colon, right dorsal colon, transverse colon, small colon, and rectum. The cecum ususally contains about 10 to 12 percent of the contents of the gastrointestinal tract, whereas the colon may contain 50 percent or more. The large intestine contains large populations of bacteria. The bacteria are similar to those found in the rumen of cattle. Although the horse cannot digest fiber, the bacteria enable the horse to utilize roughages because they produce enzymes that digest fiber. The end-products of fiber digestion by bacteria are volatile fatty acids (primarily acetate, propionate, and butyrate). These are absored from the large intestine and utilized by the horse as energy sources. A horse fed only hay may obtain one third or more of his energy from volatile fatty acids.

Bacteria are also involved with nitrogen metabolism, but the importance of that role in horse nutrition is still being studied. In ruminants such as cattle and sheep the bacteria in the rumen can utilize non-protein nitrogen sources such as urea to synthesize bacterial protein. The bacterial protein then passes to the small intestine, where it is digested and the amino acids are absorbed. However, the bacteria in the horse are beyond the small intestine. They are thus not as efficiently utilized as by the ruminant, but the extent of utilization is unclear.

It is clear that protein must be provided in the diet of young horses, but it appears that non-protein nitrogen such as urea can be utilized to improve nitrogen nutrition of mature horses that are fed low-protein diets. The urea has two possible routes: It can be incorporated into bacterial protein, and the bacterial protein is then digested and the amino acids absorbed; or the urea can be hydrolyzed into ammonia and carbon dioxide. The ammonia can be absorbed and utilized in the liver for the synthesis of certain amino

Table 2.1.   Vitamin Content of Feeds and Digesta
Demonstrating Synthesis of Vitamins by Bacteria

| | MICROGRAMS PER GRAM OF DRY WEIGHT | | | |
Sample	Niacin	Pantothenic Acid	Riboflavin	Thiamin
Hay	31	4	17	.4
Cecum	92	20	14	4
Colon	143	31	15	17
Feces	206	40	100	9

Linerode, P.A. *Proc. Amer. Assoc. Equine Pract.*, 1967, 283–314.

acids. But further studies are needed to quantify the relative importance of the two routes.

Bacteria synthesize many B vitamins (water-soluble) such as riboflavin, thiamin, and pantothenic acid. In fact, the B vitamin content of horse feces is much greater than the B vitamin content of feed (see Table 2.1). Apparently some of the B vitamins can be utilized, but the efficiency of utilization is probably not great because B vitamin deficiencies can be produced in the horse by feeding diets containing low levels of vitamins.

The relative importance of the cecum and colon is being studied at several experiment stations. In 1979 Meyer and co-workers reported that the removal of the cecum from a pony reduced the digestibility of organic matter by about 5 percent and that of fiber by 7 percent and increased water content in the feces, but that the cecum was non-essential for life. Sauer and co-workers in 1979 reported similar results. They concluded that the major part of bacterial digestion in the large intestine may well take place in the colon.

Digestibility of Feeds

Efficiency of digestion can be influenced by several factors. Some grains such as wheat, rye, and grain sorghum should be rolled, cracked, or crimped to allow the enzymes to digest them effectively. Crimping of oats improves digestibility of dry matter by about 7 percent and therefore may not be economical for horses with sound teeth. Corn can be fed cracked, whole, or on the cob.

Cooking or steaming of wheat bran did not improve digestibility in studies at Cornell. Morrison (*Feeds and Feeding*) concluded that cooking of grains does not improve digestibility. Further studies are needed to determine the nutritive value of more complicated procedures such as micronizing and popping. Potential benefits of such procedures are likely to be the result of changing starch structure and improving utilization of absorbed end-products rather than improved digestibility.

We found no improvement in digestibility when hay was autoclaved. Pelleting of hay decreases fiber digestibility because it increases rate of passage. Treating fibrous feeds such as straw with chemicals can improve digestibility, but at present it is not a practical on-the-farm procedure.

Levels of Intake

High levels of intake may decrease digestibility because the rate of passage is increased and the bacteria do not have as much time to act on the fiber.

Activity

The level of activity of the horse may influence digestibility. We found that light exercise decreased gut motility and rate of passage of food through the digestive tract and improved digestibility slightly. On the other hand, severe work could decrease digestibility.

There can be large differences among horses in their ability to digest food. However, differences in metabolism rather than in digestive ability probably account for the major differences between "easy-keepers" and "hard-doers."

Feeding Frequency

Frequency of feeding does not appear to influence digestibility if horses are fed equal amounts of feed. However if increased feeding frequency is accompanied with increased intake, digestibility of fiber may be decreased.

Enzymes

The addition of enzymes to feed is sometimes advocated as a means of improving digestibility. It is true that if a feed is incubated with the proper enzymes, the fiber could be predigested and its utilization by the horse improved. However, it seems unlikely that sprinkling a dry supplement on feed just prior to ingestion by the horse would be of much value in the improvement of digestibility. One company claimed that if their product was used there would be almost no manure to remove and labor would be greatly decreased. We tried the product with four horses and found no improvement in digestibility. (And we had to clean up as much manure as before.)

Watering

Several current books on horse care and feeding perpetuate the myth that watering a horse after eating washes feed out of the stomach and decreases digestibility. Some myths die hard. In 1891, J.W. Sanborn at the Utah Agricultural College studied the effect of time of watering horses, and found it had no effect on digestive efficiency before or after feeding. Several studies in Europe in the early 1900s produced similar results. We conducted digestion trials with ponies and also found that watering before or after eating did not influence digestion. But we found that some horses would not eat readily unless first given water; Dr. Sanborn, too, concluded that horses watered before feeding had a better appetite and ate more than those watered only after eating. It also seems logical that some horses would be thirsty after eating dry hay and grain.

When we see horses galloping across the silver screen in the great Westerns—romanticizing a gun fight at the O.K. Corral or the exploits of Billy the Kid—we don't realize that scientists at agricultural experiment stations and colleges were, at the same time, busily doing their bit behind the scenes to open up the West.

PROBLEMS OF THE DIGESTIVE TRACT RELATED TO FEEDING

Colic describes pain in the abdomen. Most types of colic are characterized by increased heart rate and respiration, sweating, lying down and rolling, or kicking at the belly. The problem can be found in any part of the gastrointestinal tract, and can be caused by a wide variety of factors. In order for a treatment to be effective, the cause of the colic must be determined first. Because of the high incidence and multiple causes of colic, mythical treatments abound. As early as A.D. 50, Columella reported that the sight of a duck swimming helped cure colic in horses and mules.

Parasites are generally considered the most common cause of colic. Many veterinarians claim that if a farm has a good parasite control program the incidence of colic is low—with or without ducks.

Strongylus vulgaris normally occurs in the large intestine. Its eggs are excreted in the feces and, if the larval stage is ingested by the horse, the larvae can penetrate the intestinal wall and end up in the mesenteric arteries (the blood vessels supplying the intestine). The larvae can be enveloped in a thrombus (clot). As the thrombus grows an aneurysm or dilation of the blood vessel may develop and the nerve supply can be impaired. The lack of adequate blood and nerve supply can cause decreased motility of the intestine, resulting in necrosis or death of a segment of the intestine. Thus parasites can cause a range of problems: from decreased utilization of food, decreased rate of growth, weight loss, or colic to death—depending on the extent of the damage. Obviously an effective parasite control program is essential for the successful rearing of horses.

Few controlled studies have been conducted on the effect of diet on internal parasites. It is a known fact, however, that horses should not be given feed that is contaminated with feces because of the danger of ingestion of eggs. Therefore, never feed on the ground in paddocks heavily populated with horses. Also, a poorly fed horse is more susceptible to the pathogenesis of parasite invasion than a properly fed horse.

C. E. Howell and M. A. Stewart (1940) reported that horses fed 17 pounds of prunes, 3 pounds of barley, and some meadow hay were more susceptible to strongyle infections than horses fed a mixture containing 30 percent barley, 30 percent oats, 30 percent wheat bran, 5 percent linseed meal, and 5 percent cottonseed meal (with no hay), or a ration containing 55 percent beet pulp, 10 percent barley, 10 percent oats, 10 percent wheat bran, 10 percent linseed meal, and 5 percent cottonseed meal—and again, no hay. Prunes appear to be the culprit. The diet was tried because several horse owners who were also producing prunes had been feeding cull prunes to their horses but found many worms in the feces. The excess sugar in the prune diet was believed to have stimulated the parasite.

Colic can be caused by overeating and distension of the stomach. As mentioned earlier, the horse seldom if ever vomits and the stomach is relatively small. Thus, greedy horses are as susceptible to colic as they are to choke. There are several ways of preventing this. Feed your horse several times a day; feed hay prior to feeding grain; spread the grain thinly in a large manger; put large, smooth stones in the manger so that the horse has to work a bit to find the food; or mix the grain with chopped hay.

Colic caused by food impaction in the large intestine may occur because of lack of adequate water intake. This type of colic is more frequently seen in the winter in northern areas because watering of horses is more likely to be neglected then.

Blockage of the opening from the stomach to the small intestine by bots can also cause colic.

Impaction

Some horses seem to have an insatiable hunger for the bizarre—anything from bedding to woodshavings or sawdust—causing colic and impaction.

One owner tried unsuccessfully to prevent his horses from chewing the stalls by nailing strips of carpet scraps to the wood. One of the horses developed colic. It was treated without success and eventually died. At autopsy, a three-foot strip of carpet was found blocking the intestine.

Dr. S. M. Getty and co-workers (1976) at Michigan State University recently reported another case in which a large object caused intestinal obstruction. A five-month old filly was presented to their clinic because feed intake was greatly reduced and the horse was losing weight, was lethargic, and suffered from recurring attacks of colic. The animal died shortly after admission to the clinic. At autopsy, large masses of cord from a rubber fence were found in the stomach, small intestine, and large intestine.

The authors advise horse owners to construct an inner strand of electric wire between horse and fence to prevent direct contact, or apply a sealer to the fence to make it less appetizing. Horses, especially the young, seem partial to chewing cord unraveled from rubberized fencing.

Several reports of impaction caused by strands from rubber fencing have also come from the University of Pennsylvania and Cornell Unviersity. In one particularly interesting case, a horse had not been on a farm with a rubber fence for three years; it took that long after ingestion of the material before the problem became serious.

Small objects might become the center of balls of ingesta called *phytobezoars*. These "grow" by accumulating plant material on the surface. One of our horses at Cornell developed chronic colic. Medical treatment was not successful, so surgery was performed to remove phytobezoars about 2 inches in diameter from the intestine. A small piece of wood was found in the center of each one. However, it is strange that only two of the many experimental horses have developed the phytobezoars, because wood chewing is not uncommon.

Enteroliths

Enteroliths, or stones of the intestine, occur frequently in some areas of the United States; they appear to be increasing in western and particularly the southwestern states. Dr. J. D. Wheat (1976) reported that clinical signs are mild until acute obstruction, which occurs usually when the enterolith passes from the large to the small colon. When obstruction occurs the signs are more severe than those caused by phytobezoars, and if enteroliths are not removed they may cause rupture of the colon. Dr. Wheat reported that the most common enteroliths usually consist of magnesium ammonium phosphate. There is usually a small nidus (core) of metal or other hard material in the center of the enterolith. Pins, nails, glazing points, pieces of wire, and pebbles are examples of objects found in enteroliths. The size and shape of the enterolith can vary greatly. Some can get as big as a bowling ball and weigh 17 pounds or more. Some are passed when they are small. The horse may have one or many stones; Evans et al. found more than 100 small enteroliths in one horse. When multiple stones are present they tend to be polyhedral (many-sided) rather than round, because the surfaces that are not in contact with each other grow faster. Evans et al. (1981) reported that Arabian and part-Arabian horses were more likely than the other common breeds to have enteroliths, but no reasons were given.

The relationship of feeding to the formation of enteroliths is not clearly defined. It seems quite likely the mineral content of the ration or water could be a factor in the formation of enteroliths, but further studies are needed. Some veterinarians have suggested that the enteroliths are more likely to develop when wheat bran and rye bran are fed in large amounts—perhaps because they contain high levels of magnesium and phosphate. Wheat bran contains 0.6 percent magnesium and 1.4 percent phosphorus, whereas oats contain 0.2 percent magnesium and 0.4 percent phosphorus. Well water containing a high level of magnesium was found on one ranch that had a serious problem with enteroliths. High iron intake has also been suggested as a causative factor. Further studies are needed to establish the role of minerals in this disorder.

Sand Colic

Horses fed on the ground in areas where the soil is sandy may accumulate large amounts of sand in the large intestine. A veter-

Three enteroliths taken from the large intestines of horses. (The softball is included merely for size comparison.) A section was cut from the enterolith on the left to illustrate the layers. A nail is visible in the center of the enterolith in the middle. The large one on the right weighed about 8 pounds. (Photograph by the author)

inarian in Arizona reported that, of the 1,042 cases of colic seen in his practice over a 5-year period, 324 were cases of sand colic.

Sand colic can often be treated, but some horses are not so lucky. Dr. T. Udenberg (1979) reported the case of a sixteen-year old Quarter Horse kept in a well-worked sand arena and fed on the ground. The horse did not respond to treatment and underwent surgery. As the colon was found to be greatly dilated with sand and a significant area of the large intestine had become necrotic, the horse had to be euthanized. Sixty pounds of sand were removed from the colon!

Horses can also eat gravel. One horse referred to the clinic at Cornell had five gallons of gravel removed from the large intestine. Gravel used in the construction of a new barn had been piled in the horse pasture.

Colitis X

Colitis X is often fatal. It is characterized by the sudden onset of profuse, watery diarrhea and rapidly developing signs of shock. Treatment is massive parenteral fluid therapy. Dr. H. B. Schiefer (1981) suggested that the onset of the disease often occurs after horses are abruptly changed to a high protein–low fiber diet. I suspect that the change in type and amout of carbohydrate reaching the large intestine is more important as a causative factor than the change in protein level, as this could affect the balance of the bacteria in the large intestine. But colitis X has also been reported in horses that have *not* had a sudden change in feed.

Enterotoxemia

Enterotoxemia (overeating disease) is caused by toxins produced by *Clostridium perfringens* Type D bacteria. Dr. T. W. Swerczek (1976) reported that the condition is most often found in the largest and fastest-growing foals in group feeding situations. Clinical signs are rarely observed. The foals appear healthy at feeding time but may be dead shortly afterward. Dr. Swerczek reported that 28 of 935 foals necropsied at the University of Kentucky had died from enterotoxemia. The gastrointestinal tract was filled with grain and dilated because of gas formation. The ingesta had a strong fermentative odor. The cortex of the kidneys was degenerative and contained areas of hemorrhage. Similar signs are found in lambs with enterotoxemia.

Sound management can help prevent enterotoxemia. Hungry foals should not be given unlimited access to grain. During group feeding, all animals should be given ample space and opportunity to eat. "Boss" foals should not be allowed to crowd out the other foals and eat more than their share. Pelleted feeds containing only grains, with no significant level of fiber, should not be fed free choice. Changes in feed should be made gradually, especially with recently weaned foals brought in from pasture and introduced to grain.

Table 2.2. Water Content of Ingesta
from Various Sections
of the Gastrointestinal Tract

Site	Percent Water
Stomach	85
Small intestine	92
Cecum	90
Colon	85
Rectum	75

Water Absorption and Diarrhea

The water content of material in the intestine decreases as the material goes from small intestine to rectum (Table 2.2).

The daily volume of water entering the large intestine is equal to the animal's total extracellular fluid volume (about 5 gallons for a pony, 15 gallons for a horse), and 95 percent of that water is reabsorbed. When the water isn't reabsorbed, diarrhea results.

Wheat bran is often advocated as a laxative feed. Although bran may increase the water content of the feces because of the fiber (Table 2.3), the digestive tract adapts to wheat bran. Also, the effectiveness of bran is minimized if the ration previously contained fiber sources. Low levels of fiber *decrease* the water content of the feces. If the feces of ponies fed only oats contained only 50 percent water, the addition of bran to such a diet would increase water content.

The feces of horses fed a complete pelleted ration or only grass will usually contain more water than feces of horses fed hay and grain. The increased water content is probably due to the faster rate of passage through the tract, allowing less time for water reabsorption.

Fonnesbeck (1968) determined the water content of the feces of horses fed several different kinds of grass hays (bromegrass, bermudagrass, canarygrass, fescue, or timothy), legume hays (al-

Table 2.3.   Effect of Diet on Water Content of Feces of Ponies
and Horses

Diet	Water Content (%)	Animals
Wet bran: 50% bran—50% beet pulp[a]	68.2	Ponies
Dry bran: 50% bran—50% beet pulp[a]	68.5	Ponies
Timothy hay and corn[a]	66.4	Ponies
Timothy hay, corn, and 16% wheat bran[a]	69.4	Ponies
Alfalfa pellets[a]	68.8	Ponies
Oats[a]	50.7	Ponies
Medium potassium (.9%)[a]	69.0	Ponies
High potassium (3.4%)[a]	71.5	Ponies
Medium protein (14.5%)[a]	76.5	Horses
High protein (22%)[a]	76.3	Horses
Grass hay[b]	76.2	Horses
Legume hay[b]	75.3	Horses
Corn and hay[b]	76.7	Horses
Oats and hay[b]	70.9	Horses

[a] Cornell studies.
[b] Fonnesbeck, P.V. *J. Animal Sci.*, 27: 1350, 1968.

falfa or clover), and combinations of hays and grains (oats and corn). The results are summarized in Table 2.3. Animals accustomed to legume hay did not have a higher water content in the feces than those fed grass hay. Oats caused a drier feces. The horses fed alfalfa and clover excreted significantly more urine than the horses fed grass hay.

Feeding high levels of minerals can also increase the water content of the feces (Table 2.3). High levels of protein (soybean meal) increased urinary excretion but did not influence the water content of feces.

The feces of horses may contain slightly more water than the feces of ponies. Donkey feces may be drier than pony feces when the animals are fed the same diets.

Diarrhea or disruption of water reabsorption can be caused by many factors such as enteritis (infection of the intestinal tract), excitement, allergic reactions, and incorrect feeding practices. A horse accustomed to being fed timothy hay and oats that is abruptly changed to alfalfa and corn is likely to have more water in the feces than usual, until the animal adjusts. A changeover to the new diet should take at least four to five days, and should incorporate 20 to 25 percent more of the new feed each day.

Diarrhea can be caused by the excessive intake of milk by foals and mature horses. Lactose (milk sugar) is broken down by the enzyme lactase, which is found in the small intestine but not in the large intestine of young horses. The activity of the enzyme decreases with age; horses over three have little or no lactase activity. Feeding high levels of milk at one time to a young horse may cause diarrhea because the enzymes cannot digest the food rapidly enough. The dam's ration can be reduced to decrease milk production, or she can be partially milked out by hand if excess milk is causing diarrhea.

An older horse may have trouble because it has no lactase. Dr. M. C. Roberts (1975) suggested that mature horses could not be fed more than .4 gm of lactose per pound of body weight without a high risk of diarrhea, so don't feed a 1,000-pound horse more than 400 gm of lactose. Dried skim milk contains about 40 percent lactose, so a horse fed more than 1 kg (2.2 lb) of dried skim milk daily would probably have a digestive upset and diarrhea.

References

Evans, D.R. et al. Diagnosis and treatment of enterolithiasis in equidae. *Comp. Cont. Educ. Pract. Vet.* S. 383, 1981.

Getty, S.M. et al. Rubberized fencing as a gastrointestinal obstruction in a young horse. *Vet. Med. Small Anim. Clinic* 71:221, 1976.

Hintz, H.F. Digestive physiology of the horse. *J. S. Afr. Vet. Assoc.* 46:13, 1975.

Howell, C.E. and M.A. Stewart. Preliminary studies on the effects of diet upon internal parasites in horses. *Amer. J. Vet. Res.* 1:58, 1940.

Meyer, H. et al. Untersuchunger uber die verdaulichkeit and vertraglichkeit verschiedener futtermittel bei typhliktamier Ponys. *Deutsche Tierarzt woch.* 86:384, 1979.

Rebhun, W., S.G. Dill, and H.T. Power. Gastric ulcers in foals. *J. Amer. Vet. Med. Assoc.* 180:404, 1982.

Roberts, M.C. Carbohydrate digestion and absorption in the equine small intestine. *J. S. Afr. Vet. Assoc.* 46:19, 1975.

Sanborn, J.W. Time of watering horses. *Utah Agr. Exp. Stat. Bull.* no. 9, 1891.

Sauer, W.S. et al. Effect of cecectomy on digestibility coefficients and nitrogen balance in ponies. *Can. J. Anim. Sci.* 59:145, 1979.

Schiefer, H.B. Equine colitis X—Still an enigma? *Can. Vet. J.* 22:162, 1981.

Swerczek, T.W. Enterotoxemia. *Proc. Soc. Theriogenology* Lexington, Ky., 1976.

Udenberg, T. Equine colic associated with sand impaction. *Can. Vet. J.* 20:269, 1979.

Wheat, J.D. Causes of colic. *J. S. Afr. Vet. Assoc.* 46:95, 1976.

Chapter Three

Nutrients

Energy

Energy is supplied by carbohydrate, protein, and fat.

CARBOHYDRATE

Carbohydrate is the body's primary source of energy. It can be divided into fractions by several different schemes. For many years the most commonly-used categories were crude fiber and nitrogen-free extract. Crude fiber remains after the feed has been treated with acid and then with alkaline solutions. Nitrogen-free extract remains after the moisture, ether extract (crude fat), crude protein, crude fiber, and ash contents are determined.

A major fault with this method is that crude fiber varies in digestibility from feedstuff to feedstuff because it is not a consistent entity. In some feeds such as soybean hulls, crude fiber is highly digestible; in other feeds such as late cut timothy, it is poorly digested.

Dr. P. J. Van Soest (1982) has developed a more precise method of analyzing feedstuffs. Boiling a feed sample with a neutral detergent solubilizes the cell contents and pectin; this leaves behind the cell wall, called neutral detergent fiber (NDF), containing cellulose, hemicellulose, and lignin.

Boiling with an acid detergent hydrolyzes the hemicellulose. This leaves behind the acid detergent fiber (ADF), which contains cellulose and lignin. Lignin content can be estimated by oxidation with potassium permanganate. Separating feeds into NDF, ADF, cellulose, hemicellulose, and lignin allows much better estimation of the nutritive value of a feedstuff than does separation into a crude fiber and nitrogen-free extract. Hemicellulose is more easily digested than cellulose. Lignin is not digestible.

FAT

Fat is a much more concentrated source of energy than carbohydrate or protein. Fat also supplies unsaturated essential fatty acids (linoleic, linolenic, and arachidonic), which are required by horses. A deficiency of fatty acids has not been produced experimentally in horses, but in other species it causes scaly skin, reduced rate of growth, and eventual death; it would probably cause similar problems in horses.

Although most horse feeds contain less than 6 percent fat, they apparently supply the essential fatty acids. However, the addition of higher levels of fat may have some benefits for hardworking horses. Increasing the unsaturated fat intake by feeding two or three tablespoons of linseed oil, corn oil, or a similar vegetable oil daily has been reported by many horse owners to improve coat quality.

PROTEIN

Protein that is not used in building body tissues can be used to supply energy. The nitrogen is excreted in the urine and the carbon chain is oxidized.

Energy Evaluation

The energy content of foods and feeds can be expressed in various ways. For humans, most energy recommendations and tables of energy content of foods—such as those used by calorie counters—

are based on the classical Atwater energy conversion factors of 4 kcal per gram of food protein and carbohydrate and 9 kcal per gram of food fat. It is assumed that the foods consumed by humans are highly digestible.

The feeds consumed by horses are not highly digestible; therefore, differences in digestibility must be taken into account. One method of evaluation is the use of total digestible nutrients (TDN). TDN is the sum of the contents of the digestible protein, digestible crude fiber, digestible nitrogen-free extract, and digestible ether extract (crude fat) times 2.25. The fat content is multiplied by 2.25 because it contains that much more energy per unit of weight. TDN has been used to measure energy content for more than 100 years and values for many feeds are available.

Another method is digestible energy, or the energy in the feed minus the energy in the feces, when the gross energy content of both is measured in a bomb calorimeter. Digestible energy values can be estimated from TDN values by assuming that 4.4 Megacalories (Mcal) are equivalent to 1 kg of TDN.

Digestible energy corrected for losses in gases and urine is metabolizable energy. Metabolizable energy corrected for heat lost in the utilization of metabolizable energy is called net energy. Net energy is the energy of the feed actually utilized by the animal to maintain life or incorporated into tissue or product. Although metabolizable energy or net energy values are used by many poultry or cattle growers because they are more precise, few such values are available for feeds for horses. Therefore digestible energy values will be used in this book.

Energy status is often easily determined. A deficiency of energy in mature horses causes loss of weight and lethargy. Mares may have delayed estrus, or fail to breed, or they may have prolonged gestation periods. Young horses that are underfed have a reduced rate of growth and the onset of sexual maturity is delayed.

Although energy status may be evident, the cause in change of status might not be so readily determined. Inadequate energy can be caused by several factors other than inadequate feed. Heavy parasite loads and dental problems are two of the most common causes of inadequate energy. An effective parasite control program is essential for the efficient use of feed. Routine dental care can improve utilization of feed. When group-feeding animals, care should be taken that all animals have adequate access to feed. Disease can increase energy needs. Some horses lose weight because they cannot digest certain components of the feed due to pathological changes in the digestive tract. Cancers such as lymphosarcoma and gastric carcinoma or infections of the tract (enteritis) decrease feed utilization. (Cancer is more likely to occur in older than younger horses.) A few horses lose weight for no apparent reason. Some spontaneously recover; others never do.

Excess feed can cause many problems such as colic, enterotoxemia, founder, and decreased life span. Other effects of overfeeding young animals and breeding animals are discussed in detail later.

Owners of show horses and pleasure horses are often accused of being primary overfeeders. As far back as 1785, Young wrote in *Annals of Agriculture* that overfeeding was a problem even when horses were used primarily for work. "What must be the expense to those farmers," he observed, "who make their fat, sleek teams an object of vanity."

A practical method of evaluating energy status is routine weighing of animals. Monthly weight records are an important management tool because appearance can be misleading. Long hair may hide weight loss. Gradual weight changes may go undetected. Chronic slow weight loss or gain may not be readily apparent to a person who observes a horse daily until the total change becomes severe. In the early stages of an illness, performance or health could be affected but the horse appears normal to its owner.

If a scale is not available, tapes placed around the heart girth can be used to estimate weight. Properly calibrated tapes are usually quite accurate for most classes of horses except pregnant mares. Several feed companies provide weight tapes free or for a nominal charge.

More precise methods of determining energy status require the determination of the amount of body fat. A horse can maintain weight but the amount of fat can be changed: A horse on a conditioning program can increase protein and water content, lose fat, but maintain weight.

Dr. Gary Potter and co-workers at Texas A & M University developed the system of visual appraisal or condition score for horses shown in Table 3.1.

Condition scores are particularly useful in the development of feeding programs for breeding animals and for conditioning horses. Fat content can also be estimated by the use of ultrasonic methods.

Estimates of energy requirements for various classes of horses are provided in Table 5.1, Chapter 5. But remember, these are only guides. Many factors influence the amount of energy needed to put an animal in the desired body condition.

Individuality is important. Some horses seem to get fat just by looking at feed; others seem to be working for the feed dealer. Cold weather also increases energy needs. Canadian workers have reported that horses in reasonable condition may require 15 to 20 percent more feed for each 10°F that the temperature falls below 30°F. Thin horses or horses with short hair may need even more.

Protein

Proteins are polymers or chains of amino acids. These vary in type and amount. Amino acids are required to form muscle, enzymes, blood cells, hormones, hoof, hair, and other tissues. Some of them can be synthesized by horse tissue. Others, called essential amino acids, must be supplied by the diet.

It is unclear which amino acids are essential for the horse. The young horse probably requires the same amino acids essential for other species such as the pig and rat. Those are listed in Table 3.2. The only amino acid studied extensively in horses is lysine; young horses require about 0.7 percent of it in the diet for rapid rate of gain. Because the requirements of other amino acids are unknown, the nitrogen requirements for the horse are expressed in terms of protein.

Table 3.1. Condition Score for Horses

Score

1   *Poor*   Animal extremely emaciated. Spinous processes, ribs, tailhead, and hooks and pins projecting prominently. Bone structure of withers, shoulders, and neck easily noticeable. No fatty tissues can be felt.

2   *Very Thin.*   Animal emaciated. Slight fat covering over base of spinous processes; transverse processes of lumbar vertebrae feel rounder. Spinous processes, ribs, tailhead, and hooks and pins prominent. Withers, shoulders, and neck structures faintly discernable.

3   *Thin*   Fat built up about halfway on spinous processes; transverse processes cannot be felt. Slight fat cover over ribs. Spinous processes and ribs easily discernable. Tailhead prominent, but individual vertebrae cannot be visually identified. Hook bones appear rounded but easily discernable. Pin bones not distinguishable. Withers, shoulders, and neck accentuated.

4   *Moderately Thin*   Negative crease along back. Faint outline of ribs discernable. Tailhead prominence depends on conformation; fat can be felt around it. Hook bones not discernable. Withers, shoulders, and neck not obviously thin.

5   *Moderate*   Back level. Ribs cannot be visually distinguished but can be easily felt. Fat around tailhead beginning to feel spongy. Withers appear rounded over spinous processes. Shoulders and neck blend smoothly into body.

6   *Moderate to Fleshy*   Slight crease down back. Fat over ribs feels spongy. Fat around tailhead feels soft. Fat beginning to be deposited along the sides of the withers, behind the shoulders, and along the sides of the neck.

7   *Fleshy*   Crease down back. Individual ribs can be felt, but noticeable filling between ribs with fat. Fat around tailhead is soft. Fat deposited along withers, behind shoulders, and along the neck.

8   *Fat*   Prominent crease down back. Difficult to feel ribs. Fat around tailhead very soft. Area along withers filled with fat. Area behind shoulder filled in flush. Noticeable thickening of neck. Fat deposited along inner buttocks.

9   *Extremely Fat*   Extremely obvious crease down back. Patchy fat appearing over ribs. Bulging fat around tailhead, along withers, behind shoulders, and along neck. Fat along inner buttocks may rub together. Flank filled in flush.

Developed by Dr. G. Potter and co-workers at Texas A & M University, 1981.

Table 3.2.   Classification of Amino Acids with Respect
to Their Growth Effects in Rats

Essential	Nonessential
Lysine	Glycine
Tryptophan	Alanine
Histidine	Serine
Phenylalanine	Cystine*
Leucine	Tyrosine**
Isoleucine	Aspartic acid
Threonine	Glutamic acid
Methionine	Proline
Valine	Hydroxyproline
Arginine	Citrulline

*Cystine can replace about one sixth
of the methionine requirement but has
no growth effect in the absence of
methionine.
**Tyrosine can replace about one half
of the phenylalanine requirement but
has no growth effect in the absence
of phenylalanine.

PROTEIN DEFICIENCY

A deficiency of amino acids causes a marked decrease in feed
intake and reduced rate of gain. In one of our studies, foals fed a
ration containing 9 percent protein ate about two thirds of the
amount of feed consumed by foals fed diets containing 14 percent
protein. After $4^1/_2$ months of being fed the deficient ration the foals
had an average total gain of 5 pounds. The foals fed 14 percent
protein had an average total gain of 200 pounds during the same
period. The addition of protein to the deficient diet increased intake; by the twelfth month of the trial, no differences in body weight
existed between the two groups.

Protein deficiency in mature animals may cause weight loss.

Deficiency in mares may result in reproductive dysfunction and small foals.

Protein-deficient animals may have rough hair coats. L. M. Slade (1980) suggested that hair follicle bulb length, bulb widths, hair shaft diameter, and hair growth rate may be decreased in protein-deficient animals.

For severely deficient animals the plasma protein and albumin levels may be decreased. The horses in the above studies fed the diet with 9 percent protein had plasma protein and albumin levels of 5.6 and 3.5 mg per 100 ml respectively, whereas the foals fed the 14 percent protein diet had values of 6.3 and 4.11 mg per 100 ml respectively. L. M. Slade (1980) suggested that in protein-deficient horses the ratio of albumin to globulin is decreased.

When high levels of protein are fed, the excess nitrogen is excreted in the urine. Thus the feeding of excess protein is wasteful, as protein is usually more expensive than carbohydrate. But there is no evidence to suggest that extra protein causes kidney damage.

High protein intakes cause increased urinary calcium losses in humans and rats, but we were unable to find similar effects in horses fed 20 percent protein.

UREA UTILIZATION

Ruminants (cattle and sheep) are often fed nonprotein nitrogen sources such as urea. Bacteria in the rumen convert the nitrogen to bacterial protein, which the animal digests in the small intestine. Mature horses can use some nonprotein nitrogen to increase nitrogen supplies. But it is not clear whether the nonprotein nitrogen is utilized by the horse tissue to synthesize nonessential amino acids, or if bacteria of the cecum or colon use the nitrogen to form protein that is subsequently digested, and from which amino acids are absorbed from the large intestine.

Workers at Colorado State University reported that mares fed low protein rangeland grass would benefit from having access to a protein block of 20 percent crude protein which included 2.13 percent urea. They concluded that under Western range conditions brood mares may be fed protein blocks in a management system to minimize the amount of time and facilities required to supplement them.

The horse does not utilize nonprotein nitrogen sources as efficiently as do ruminants. Although nonprotein nitrogen may have some value for mature horses, it is not recommended for young horses.

Horses are less sensitive to overfeeding of urea and other nonprotein nitrogen than ruminants. When ruminants are fed excessive amounts of these compounds, the bacteria break down the compounds into ammonia; nevertheless, the concentration is too great for the bacteria to use it all to form protein. So ammonia is absorbed across the rumen wall into the blood, and high levels of blood ammonia can be toxic. The horse is more tolerant of high levels of urea than the cow. Much of the urea fed to a horse is absorbed from the digestive tract and excreted in the urine before the urea reaches the large intestine, where it can be converted by bacteria into ammonia and carbon dioxide.

Urea dietary levels as high as 5 percent have been fed to horses without apparent difficulty. However, ponies weighing 285 pounds fed one pound of urea (about 25 percent of the diet) died of ammonia toxicosis. The first signs were aimless wandering and incoordination, followed by the pony pressing its head against fixed objects. Once this started it usually continued until the animal fell and died shortly afterward. The first signs were manifested 2 to 10 hours after the urea was fed, and death usually occurred 30 to 90 minutes later.

We are frequently asked if a dairy cattle feed containing urea will hurt a horse. If the level of urea in the feed is safe for cattle, it will be safe for horses. But bear in mind that other factors in the feed may cause problems.

Minerals

At least 21 mineral elements are required in the diets of horses. Seven—calcium, phosphorus, magnesium, sodium, chloride, potassium, and sulfur—are considered major elements, not because they are any more important but because they are required in larger amounts. The others are called trace minerals because the amount required is quite small. The list of required trace minerals is increasing steadily. Chromium, nickel, silicon, vanadium, and tin are examples of minerals that have been only recently considered as required, and others will probably follow because of new analytical procedures and experimental methods. Usually the minerals recently recognized as essential, such as tin, are not of practical concern. In fact, it is very difficult to produce a deficiency of these substances; therefore, they will not be discussed further in this book. However, future developments may change their practical importance in animal feeding.

The amounts of minerals required in animals' diets can be influenced by many factors. The function of the animal influences need. Growing animals, lactating mares, or pregnant mares require greater levels than do mature animals. Genetic variations can also influence requirements. The genetic factor has not been studied in horses, but many examples can be given for other species. Certain lines of Malamutes have impaired ability to absorb copper, whereas lines of Bedlington Terriers are very sensitive to copper diets. Thus diets that contain a level of copper that is adequate and safe for most dogs can produce a deficiency in some Malamutes or liver damage due to excess in some Bedlington Terriers.

One of the primary factors influencing the concentration of minerals needed in the diet is bioavailability, or the efficiency with which animals absorb and utilize the minerals. The chemical and physical form can greatly influence bioavailability. Unfortunately, most of the experimental work on bioavailability has been done with rats and chicks and is not strictly applicable to horses, but some inferences can be made. Iron compounds in the ferrous states—such as ferrous sulfate, ferrous chloride, ferrous ammonium sulfate—are readily utilized by rats, pigs, and chicks; but ferric compounds such as ferric chloride are used much less efficiently. Ferric oxide, which is really rust, is almost totally unavailable. However, ferric oxide is added to some mineral mixtures and dog foods to provide color.

The horse owner seldom buys individual trace minerals. He relies on the feed manufacturers to select the best form based on bioavailability, price, supply dependability, and handling properties.

The content of other minerals or ingredients in the diet can also influence bioavailability. High levels of calcium, for example, can decrease the utilization of zinc. High levels of phosphorus can decrease calcium bioavailability. Minerals should thus be kept in balance. Mineral bioavailability and content of feeds is influenced by soil mineral content, plant species, stage of maturity, and methods of harvesting.

Because trace minerals and vitamins are required in such small amounts, the requirements are usually expressed in metric units. The most common unit is milligrams per kilogram, which is expressed as parts per million (ppm). One ppm is one milligram per million milligrams (kilogram); a penny is one ppm of $10,000, or the proverbial one-gram needle is one ppm in 2,200 pounds of haystack.

CALCIUM AND PHOSPHORUS

Calcium and phosphorus have important interrelationships and both are required for the mineralization of bone.

Bone development is a process usually associated with young animals during growth, However, demands are made on the mineral reserve of bone throughout life—particularly on calcium and phosphorus during reproduction and lactation.

Bone is a dynamic tissue; it is continually undergoing change. Nutrient deficiencies may result in serious bone problems at any time during the life of an animal. But proper nutrition alone will not guarantee a sound skeleton unless combined with proper care. Bone and joint problems are among the most common and serious causes of debility in animals; they may result from poor nutrition and the stresses of overwork. Many factors such as age, genetic and conformational faults, overuse, improper training, and working surface can cause skeletal problems in horses fed excellent diets.

Of course, apart from sound care, proper bone formation requires many nutritients in addition to calcium and phosphorus. These include vitamins A and D, energy, protein, and the trace

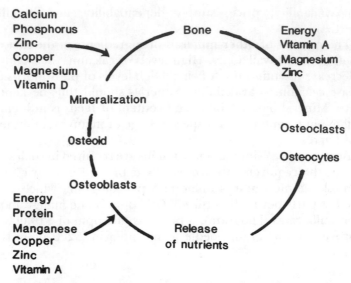

Figure B. **Relationship of nutrients to bone formation.**

minerals copper, zinc, and manganese. The general relationship of these nutrients to bone formation and bone resorption is shown below in Figure B.

A deficiency of either calcium or phosphorus will result in reduced mineralization of the organic framework of bone. The bones will be weak; long bones in the legs may become bowed under the pressure of the weight of the animal. The ends of the bones may appear enlarged. A simple deficiency is called rickets in young animals or osteomalacia (soft bone) in older animals.

Nutritional secondary hyperparathyroidism results if horses are fed a diet low in calcium and adequate or higher in phosphorus. Calcium has several vital functions in addition to bone formation. An adequate blood level of calcium is required for normal muscular activity, blood clotting, release of hormones, and enzyme activation. If the blood level of calcium is too low, the animal may show unsteadiness when walking; it may also drop to the ground and not be able to get up. The eyes appear dull. If not treated, the animal may die.

The most commonly seen condition of low blood calcium in livestock is milk fever in dairy cattle, when an excess of calcium is released in the milk and is not replenished from calcium stored in the bone. A similar condition can occur in lactating mares. It is, however, much less common than milk fever in cattle and is called eclampsia.

Animals have their own regulatory system. Whenever the blood calcium drops, the parathyroid gland is triggered to release a hormone called parathormone. This hormone brings about the release of calcium from bone to restore the normal blood level of calcium.

When a horse is fed a diet containing low levels of calcium for long periods, considerable amounts of calcium can be removed from all parts of the skeleton. Lameness results because the cortex of the bone becomes thin and weak and does not provide adequate support for tendons and ligaments. Most affected horses develop a peculiar gait that resembles a rabbit hopping.

The head of an affected horse may have enlargements on the upper or lower jaw as fibrous connective tissue invades the area when calcium is removed. Hence a common name for nutritional secondary hyperparathyroidism is "big head disease." Another old name for the disease is "miller's disease"; when millers were using horses for power, they frequently fed them large amounts of milling by-products such as wheat bran, containing low levels of calcium and high levels of phosphorus.

Mules apparently are less susceptible to "big head" than horses, but we don't know why. A. W. Bitting reported in 1894 that big head was less of a problem in mules than in horses in Florida. In 1932 Kinter and Holt reported that in the Philippine Islands the incidence of big head was about three times higher in horses than in mules fed similar diets.

Fortunately, severe cases of nutritional secondary hyperthyroidism are not common today. A few cases have been referred to the Cornell Large Animal Clinic in recent years. The cases were usually animals fed large amounts of oats or corn (which contains almost no calcium) and some grass hay, in order to induce rapid growth. In three cases, the owners were quite inexperienced with livestock. They thought they were feeding alfalfa, which contains a high level of calcium, but were really feeding timothy, which does not.

A horse with big head will usually recover from the lameness after the addition of calcium to the diet. The enlargement of the head may or may not remain, depending on size and the time span of the affliction.

Although severe cases of calcium deficiency are not common, subclinical or marginal calcium deficiency may be a much more prevalent problem. However, it is doubtful that 90 percent of all horses suffer from calcium deficiency, as reported by some manufacturers of supplements.

How can you tell if horses are being fed adequate amounts of calcium? Blood samples are usually of no value because normal values are maintained by using calcium removed from the bone even when the animals are fed a deficient diet. Some companies have advertised that they can determine the calcium and phosphorus levels of the horse by hair analysis. But we have not found this to be reliable, since several factors are likely to influence calcium and phosphorus content of hair. The rate of hair growth, season of the year, and hair color have greater influence on hair content than the content of the diet. (Dark hair has a greater calcium content than white hair. Based on hair analysis, the horse owner would have to feed the white part of the pinto differently than the black part.)

Radiographs can be helpful, but early changes in the bone might go undetected by routine radiography. Some reports indicate that more than 30 percent of the calcium must be removed from bone before the changes can be detected by routine radiography.

Urinalysis can be helpful, because the amount of calcium excreted is somewhat dependent on intake. A calcium-deficient animal would excrete little calcium in the urine.

Ration analysis and comparison to estimated requirements is one of the most useful methods of evaluating calcium and phosphorus status. However, even ration analysis is not foolproof. Nutritional secondary hyperparathyroidism has often been reported in horses in tropical countries. Their rations appear to contain calcium, but several tropical grasses (such as *Setaria*) contain high levels of oxalic acid, which decreases the bioavailability of calcium

Horse with nutritional secondary hyperparathyroidism, commonly called big head disease. The upper jaw is enlarged because calcium is replaced by fibrous connective tissue. (Courtesy T. R. Joyce, Texas A & M University)

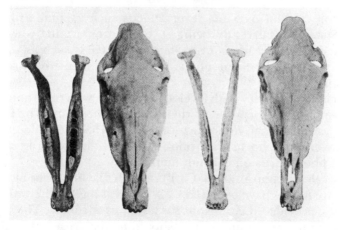

Skulls from a normal horse (right) and from a horse with nutritional secondary hyperparathyroidism (left). Note the increased width of the lower jaw bone. Although larger, the bone lacks calcium and density; fibrous connective tissue content is greatly increased. (Florida Agricultural Experimental Station Bulletin 26, 1894)

Table 3.3.   Calcium and Phosphorus Content
of Several Supplements

Supplement	Ca (%)	P (%)
Ammonium Polyphosphate	—	13
Bone meal, steamed	24	12
Diammonium phosphate	—	20
Dicalcium phosphate	23	18$^1/_2$
Limestone	35	—
Monoammonium phosphate	—	23
Monocalcium phosphate	20	21
Monosodium phosphate	—	26
Sodium tripolyphosphate	—	25
Tricalcium phosphate	38	18

by forming calcium oxalate. Several feeds such as alfalfa in the United States are currently being studied to determine whether they contain harmful levels of oxalate. Some feeds such as wheat bran contain high levels of phytic acid phosphorus, which binds the calcium.

Calcium fed at very high levels can interfere with trace mineral and phosphorus absorption. If the levels of phosphorus or trace minerals are marginal, deficiency can result. However, the effect of calcium on phosphorus absorption is not as dramatic as the effect of excess phosphorus on calcium metabolism.

The calcium:phosphorus (Ca:P) ratio should always be at least 1:1. Mature horses have been fed Ca:P ratios as high as 6:1 without difficulties, provided the phosphorus level was adequate. The Ca:P ratio in the ration of young horses should not be greater than 3:1.

In spite of many claims, there is no optimal ratio of Ca to P. If the levels of the minerals are adequate, horses can easily adapt to the range of ratios discussed above.

Legumes are excellent sources of calcium; grasses are not. Grains contain almost no calcium. The phosphorus content of forage depends on the soil content, and soil in many areas of the West and Florida is deficient in phosphorus. Many inorganic sources such as those listed in Table 3.3 are efficiently utilized by horses. Raw rock phosphate should not be used because it may contain a toxic level of fluoride content. In addition, overheating can decrease the availability of phosphorus. Phosphorus in colloidal phosphate or soft rock phosphate may also have reduced bioavailability.

Animals can obtain minerals from a self-feeder—particularly if the mineral mixture contains salt. However, studies at Cornell University indicate that horses do not have nutritional wisdom for calcium. When fed a calcium-deficient diet, weanlings and yearlings did not select adequate amounts of calcium from supplements provided free-choice to meet the National Research Council recommendations. Therefore, for young animals any needed supplements should be added to feed, rather than relying on an animal's sense to select what it needs. If fed a marginally deficient diet, mature horses may obtain adequate calcium if provided free choice a mixture of 1 part limestone and 2 parts trace mineralized salt. When both calcium and phosphorus are needed, the mixture should contain 1 part limestone, 1 part dicalcium phosphate, and 1 part trace mineralized salt. The calcium and phosphorus requirements are shown in Tables 5.2 to 5.6 in Chapter 5.

Hard water may supply as much as 5 percent of the animal's calcium requirement. Interestingly, the calcium content of hard water is believed by some scientists to be the reason why fewer people develop cardiovascular disease in hard-water areas compared with soft-water areas. Dr. Elinder and co-workers (1980) in Sweden reported that horses living in soft-water areas developed microscopic changes in the aorta and myocardium twice as frequently as horses living in hard-water areas. Although the differences were not significant, they concluded that horses are well suited for studies of the "water factor."

MAGNESIUM

Magnesium is another of the minerals found primarily in bone, but it has many other functions. It is an activator of a great number of enzymes. In the plant world, magnesium is a constituent of chlorophyll and is essential for photosynthesis. It is necessary for all green plants as well as for the animals that feed on them.

Grass tetany is a common result of magnesium deficiency in cattle. Magnesium deficiency is likely to develop when cattle graze on pasture containing low levels of magnesium, or when factors such as high levels of nitrogen, potassium, or organic acids which tie up the magnesium are present. The cattle become excessively nervous, their muscles twitch, they have difficulty breathing, and they develop a rapid pulse, resulting in convulsions and possible death. Fortunately, horses grazing on the same pastures causing grass tetany in cattle seldom have difficulty, although some cases of grass tetany in horses have been reported in humid grassland areas.

Several years ago signs similar to grass tetany were reported in ponies after several hours of being transported on railroad cars. Dr. D. D. Harrington (1974) produced severe magnesium deficiencies by feeding a purified diet to growing foals. Signs included convulsions, sweating, and padding of legs. At necropsy, mineralized plaques were found in arteries, spleen, lungs, and muscle.

The level of magnesium in blood serum indicates the magnesium status of the horse. The normal serum magnesium level is 2.2 to 2.8 mg per 100 ml of serum. Harrington found concentrations as low as 0.9 mg per 100 ml in deficient foals.

The magnesium requirement for growing horses is estimated to be 0.1 percent of the diet and 0.90 percent for mature horses. Most hays and grains contain 0.1 to 0.3 percent magnesium. If a supplement is needed, magnesium oxide is inexpensive and efficiently utilized by horses.

POTASSIUM

Potassium exists in the body primarily as a cellular constituent. It is important in osmotic pressure regulation and acid-base balance in the cells. When an animal is depleted of potassium, transmission of nerve impulses is impaired and muscular paralysis develops. Heart abnormalities such as cardiac arrhythmias may also occur in potassium-deficient animals. Dr. H. D. Stowe (1971) reported that ponies fed diets containing low levels of potassium had decreased feed intake, reduced rate of gain, and hypokalemia (decreased serum level of potassium).

Forages are excellent sources of potassium, so a potassium deficiency would only be expected in horses fed high levels of grain and little hay. Hardworking horses that lose a significant amount of potassium in sweat may respond to a mixture of electrolytes containing potassium. Low levels of serum potassium have been reported in endurance horses.

LICORICE

Never feed horses large amounts of licorice.

Of course, knowledgeable horsemen are not expected to feed large amounts of licorice to a horse, but the relationship between licorice and potassium is quite interesting. Licorice contains a compound called glycyrrhizinic acid, which mimics the action of the hormone aldosterone and causes a decrease in the blood level of potassium. The low level (hypokalemia) causes muscular weakness. There are many examples of potassium problems in humans caused by licorice excesses. One woman ate 1 to 2 ounces of licorice candy daily for 9 months. She became very weak and could hardly walk, but recovered promptly when she stopped eating licorice and was given a potassium supplement. A more bizarre case involved an 85-year-old man admitted to hospital because of muscular weakness. He had a 50-year habit of chewing 8 to 12 3-ounce bags of chewing tobacco daily and swallowing the saliva. The tobacco contained 8.3 percent licorice paste. He recovered when he stopped "plug chawin' " and was given a potassium supplement!

SALT

Salt was not known to be a compound of sodium and chloride until 1700, but the importance of salt has been known for centuries. The selling of salt was one of the major enterprises in ancient times; when supply was low, salt commanded high prices. The word *salary* comes from the practice of giving Roman soldiers money to purchase salt. "A man worth his salt" is a man earning his pay.

Sodium and chloride are important in maintaining osmotic pressure and acid-base equilibrium; they also affect water metabolism. Sodium is found primarily in extracellular fluids, whereas chloride is found both within and outside the cells of body tissue.

Deficiencies of sodium and chloride have not been described under experimental conditions in horses. But in other species, rough hair coat, lack of appetite, decreased rate of growth or even loss of weight, and production losses have been reported. Milk production decreases in cows, and egg laying decreases in chickens deprived of sodium and chloride. Cannibalism has even been reported in sodium-deficient chickens.

How much salt is needed to provide the sodium and chloride required by horses? Surprisingly few controlled studies have been conducted. In an early Michigan study with draft horses given free access to loose salt, they ate about 50 grams per day; mules usually ate less than 30 grams. The average range of intake was great. Dexter, a 5-year-old Percheron gelding, ate about 10 grams per day whereas Duke, a 9-year-old Percheron-Belgian cross gelding, ate 90 grams per day—even though the conditions were similar for both horses. Other studies indicated that 60 grams of salt per day might be consumed by horses working in humid climates. The 1978 NRC committee concluded that although the requirements are not clearly established, if salt is fed at a rate of 0.5 to 1 percent of the diet or if salt is given free-choice, a deficiency is not likely to occur.

Most grazing animals have a craving for salt; voluntary intake usually increases when the apparent needs increase. We found that horses in training consumed more salt than when they were not in training, and lactating mares consumed more salt than open mares.

Because salt is a condiment, voluntary intake may greatly exceed requirements. Although the excess salt is not toxic if water is available, it is wasted and urination increases. Excess salt without water can be very dangerous. It can cause digestive disturbances (salt cramps, diarrhea, and colic), frequent urination, weakness, loss of coordination, paralysis of the hind limbs, and even death. Excessive intake is likely to occur when animals drink salt brine or when salt-starved horses are given salt free-choice without available water.

There are rare cases of horses that cannot be allowed salt free-choice. Drs. Buntain and Coffman (1981) reported that a yearling filly was brought to the clinic at the University of Missouri because of muscle tremors and a stiff gait. The filly was drinking 50 gallons of water daily! She devoured great amounts of salt when given free choice. No physiological reasons were found for the high salt intake, so they concluded that the intake was psychogenic: The filly was just crazy about salt. When the salt was removed, water intake decreased, the tremors disappeared, and her gait improved.

Salt can be provided in loose or block form. Intake of loose salt may be greater than that of the block form, but in general adequate amounts can be consumed from either form. In any case, the trace-mineralized form should be used because it provides some insurance for proper trace mineral nutrition.

Trace-mineralized salt may contain 0.2 percent manganese, 0.15 percent iron, 0.3 percent copper, 0.01 percent cobalt, 0.008 percent zinc, and 0.007 percent iodine. A mature horse weighing 1,000 pounds eating 18 pounds of hay and 30 grams of salt daily would obtain 100 percent of the copper, 100 percent of the cobalt, 1 percent of the zinc, and 250 percent of the iodine requirement. Although 30 grams would provide more iodine than the requirement, the level would not approach toxic levels. Some trace-mineralized salts (particularly those formulated for swine) may contain much higher levels of zinc than in the above example. Some custom-formulated salt mixes could also provide selenium.

SULFUR

Sulfur requirements have not been studied in horses. The estimated requirement mentioned in this text is based on studies of other species. Sulfur is present in the amino acids methionine and cystine as well as in chondroitin sulfate, a constituent of cartilage. Most rations commonly fed to horses contain levels of sulfur higher than the estimated requirement, and sulfur deficiency is not usually considered to be of practical concern.

In the early 1900s natural sulfur water was highly prized for horses; even today some horsemen feed sulfur as a tonic or because they feel additional sulfur will improve hoof and hair growth. Hoof and hair contain 2 to 3 percent sulfur and have a significant content of sulfur amino acids, but no research to support claims for sulfur supplementation is available.

Sulfur supplementation is not without danger. Dr. M. J. Corke (1981) reported that horses in England are often given sulfur as a tonic but that overfeeding can cause death. He cited one case in which flowers of sulfur was given to 14 horses; by mistake each horse was fed between one-half to almost one pound of sulfur. Within 12 hours, the horses appeared dull and lethargic, and showed signs of colic. At 48 hours stable yellow froth appeared in the nostrils. By 60 hours jaundice was noted, and petichial hemorrhage (small spots) appeared on the gums. Only two horses died, but liver dysfunction was determined in the surviving horses with possible long-term harmful effects.

Dr. R. M. Jordan (1981) reported that young ponies fed high levels of sulfur mixed with grain refused to eat and lost weight.

IRON

Iron is a constituent of the oxygen carrier hemoglobin, which is contained in red blood cells. A deficiency of iron causes anemia (lack of red blood cells).

The maintenance requirement for iron has not been studied thoroughly but is probably less than 40 ppm. For growing foals dietary levels of 50 ppm iron are adequate. Most feeds contain levels much higher than the estimated requirements. Grains may contain 200 to 400 ppm iron, and hay may contain 100 to 300 ppm iron. The body effectively conserves iron, so simple iron deficiency is not commonly reported. Chronic blood loss due to heavy parasite loads can greatly increase the loss of iron and can induce an iron deficiency.

Few studies have been conducted on iron toxicity levels for horses. However, cases have been reported of fatalities in horses injected with iron compounds. Some horses apparently have an undefined hypersensitivity to iron dextran injections and may have a fatal reaction, although such incidences are rare. One veterinarian reported that he had been using iron dextran injections for 13 years with no apparent problems.

Reported cases of toxicity caused by ingesting iron are even more rare. Dr. J. Arnbjerg (1981) reported the case of a riding horse that had not been performing as well as expected. The owner decided to give him an iron tonic but by mistake gave him more than a pound of ferrous fumarate over a five-day period. On the sixth day the horse stopped eating, lacked coordination, and was sweating although he had depressed body temperature. A few hours later the horse was in shock; he died shortly afterward.

No estimates of the toxic level of iron for horses are available. The maximum tolerable level of iron for cattle and chickens is 1,000 ppm. The toxic level is 3,000 ppm for swine and only 500 ppm for sheep, according to the National Research Council. Levels greater than the tolerable level may decrease growth rate and interfere with phosphorus metabolism, resulting in poorly mineralized bone.

SELENIUM

Selenium is one of the most unusual and interesting minerals. It was named after the Greek word for moon (*selene*) and, like the moon, selenium certainly has many phases. The toxic properties of selenium will be discussed first because toxicity was recognized as a problem in horses long before selenium was identified as a required mineral. Drs. Trelease and Beath (1949) suggested that Marco Polo in 1295 may have been referring to selenium toxicity when he noted that the eating of certain plants in western China caused horses' hooves to drop off. In 1860, an army surgeon reported that many of the U.S. Army horses grazing on land near Fort Randall in Nebraska Territory (now part of South Dakota) lost their manes, tails, and hooves and died. In fact, Dr. E. V. Wilcox (1944) claimed that Custer's fall at Little Big Horn was partially due to selenium toxicity. Major Marcus Reno's troops did not reach Custer because the pack train was slow—making about 10 miles

Horse with selenium toxicity. The hoof may be sloughed off. (Courtesy R. Perce, Colorado State University)

per day rather than the expected 30 miles. One of the drivers said many of the horses and mules were lame because of very sore feet. Some horses died and some acted like locoed animals—all signs which could be due to selenium toxicity. Reno's horses were wintered in an area with seleniferous soils. The army was short of hay and the animals were allowed to graze native forage. Thus Wilcox concluded that the horses were suffering from selenium toxicity. The horses belonging to the braves of Sitting Bull were wintered outside of the seleniferous belt.

Although heavy losses of livestock occurred in many areas of the western United States, it was not until the 1930s that selenium was identified as the toxic element. The areas in which crops contain high levels of selenium are shown in Figure C.

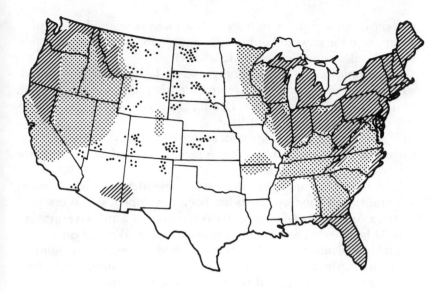

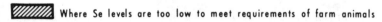 Where Se levels are too low to meet requirements of farm animals

☐ Where Se is adequate to meet requirements of farm animals

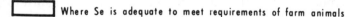 Where Se is both adequate and inadequate in same locality

• Where Se toxicity may be a problem

Figure C.   Selenium content of forages and feed crops in various areas. (From "The Effect of Soil and Fertilizers on Human and Animal Nutrition," USDA Bulletin 378, 1975)

Indicator plants such as milk vetch (*Astragalus*), prince's plume (*Stanliya*), and goldenweed (*Oonopsis*) only grow in areas where the soil is seleniferous. These plants may contain 1,000 to 10,000 ppm selenium, although 100 to 300 ppm selenium is more likely. Ingestion of such plants can cause acute selenium toxicity. The animals become depressed, breathing is labored and irregular, and death due to respiratory failure usually occurs within a few days. An estimated 150 milligrams of selenium per 100 lb of body weight can cause acute toxicity in horses. If a plant contained 4,500 ppm selenium, a horse would only need to eat $^3/_4$ lb of the plant to obtain enough selenium to cause acute toxicity.

Chronic selenium poisoning causes loss of vitality, anemia, stiffness of joints, lameness, rough hair coat, loss of mane and tail, and hoof deformities or loss. Death may occur within two months but often takes longer.

Grasses, hay, or grains containing 10 to 30 ppm selenium can cause chronic toxicity. These levels can be reached when crops are raised in seleniferous soils.

No effective manner of treating seleniferous soils to avoid selenium absorption by plants has been developed, so it is essential to fence off potentially toxic areas to limit grazing. Overgrazing should be avoided as animals are much more likely to eat the accumulator plants if other plants are limited. Arsenic is antagonistic to selenium and can be used to counteract selenium toxicity. However, great care is needed because of the toxic nature of arsenic itself.

The importance of selenium as a required mineral for livestock was first demonstrated in the 1950s, when selenium was shown to prevent white muscle disease (muscular degeneration, muscular dystrophy) in lambs and calves. Since the 1950s there have been many reports associating selenium deficiency with white muscle disease in foals. Affected foals are weak, may have difficulty nursing, and, if untreated, may die of respiratory failure. At necropsy the skeletal muscles and heart muscles are often very pale because of degeneration. The condition may appear in foals from one day of age to several months, but most often occurs within 40 days after birth.

White muscle disease has also been reported in mature horses but is considered unusual. However, other selenium deficiency problems have been suggested in mature horses. Selenium treatment is often used for azoturia or tying-up disease. Reports from England have claimed that a low performance among racehorses was related to low selenium levels. Reports from New Zealand have indicated that fertility increased when mares on farms in selenium-deficient areas were treated with selenium.

The blood selenium and vitamin E content were recently surveyed on four breeding farms in New York by Maylin and co-workers. The mean blood selenium concentration was 7.7 μg/100 ml when horses were given local feed and 15.6 μg/100 ml when fed commercial feed. Both averages were within the normal range. But on a farm that had a case of white muscle disease in a foal, the average value in the mares was only 4.2 μg/100 ml. Those mares were given 1 mg of selenium and 200 IU of vitamin E daily for eleven weeks, and the blood selenium level increased to 12.3 μg/100 ml. Mares in selenium-deficient areas fed only local feeds should therefore be given selenium supplementation.

Selenium can be given by injection or added to feed. However, don't overdo it; follow instructions carefully because of the danger of excessive selenium. The supplementation should be added to the feed so that the final concentration in the total ration is about .1 ppm of selenium. The supplement must be mixed into the feed thoroughly. Fortunately, several feed manufacturers now add selenium to feed to be sold in selenium-marginal or -deficient areas.

The average linseed meal might be expected to contain more selenium than the average soybean meal. This is because much of the flax grown in the United States comes from areas that have at least moderate levels of selenium in the soil. Studies at Cornell demonstrated that the addition of linseed meal containing 1.1 ppm selenium (or about 10 times the required level) to certain sheep rations, decreased the incidence of white muscle disease in lambs. Linseed meal might therefore be expected to supply selenium to horses.

Selenium status of animals can be evaluated by determining blood selenium levels, but this is an expensive process. It can also be evaluated by determining the level of the enzyme glutathione peroxidase in blood. Selenium is an essential part of glutathione peroxidase, which decreases intercellular peroxide. When elevated peroxide levels are present, they react and destroy parts of the cell.

Horses in areas where white muscle disease in foals had occurred had low activities (less than 20 units) of the enzyme, so the enzyme assay seems to be a sound indicator of selenium status. It is much cheaper than selenium assay, and rapid-screening blood-spot tests for the enzyme have been used with cattle and sheep.

Development of the test for horses could result in a quick and inexpensive method of determining the selenium status of horses.

Selenium and vitamin E may both be involved in the maintenance of normal muscle; they should both be included in the diet. They can also act synergistically in the prevention of white muscle disease.

IODINE

Iodine is essential for reproduction and normal physiological processes. It is a component of the hormone thyroxine, which is produced in the thyroid gland. Thyroxine and related hormones influence practically every organ in the body. They influence metabolic rate and oxygen consumption, increase uptake and utilization of glucose by the cells, and increase protein synthesis. Iodine deficiency may result in an enlarged thyroid gland (goiter) because the gland is stimulated by another hormone whenever the thyroxine blood level drops.

Feeding an iodine-deficient diet to mares can result in very weak or stillborn foals. The weak foals have difficulty standing, labored breathing, and a rapid pulse rate. The thyroid gland may or may not be enlarged. Some stillborn foals have been hairless. Iodine-deficient mares may exhibit abnormal estrus cycles but may not have goiter even though the foals do.

Soils in many widely distributed areas of the world—especially in high, mountainous regions—are very low in iodine. Sections of Wisconsin, Ohio, Iowa, Indiana, Illinois, Michigan, Montana, Nebraska, New York, North and South Dakota, Utah, Nevada, Colorado, Oregon, California, and Washington all have iodine-deficient areas.

Iodine can be effectively supplied by feeding trace-mineralized or iodized salt free-choice. American Feed Control officials state that any salt claimed to be iodized must contain not less than 0.007 percent iodine.

Signs of iodine deficiency in humans and livestock have been observed since antiquity. The use of iodized salt has greatly decreased the incidence of iodine deficiency, but sporadic cases continue to occur. Drs. Doige and McLaughlin (1981) reported that seven foals submitted for necropsy to the Western College of Veterinary Medicine at Saskatoon, Saskatchewan, from 1974 to 1980 apparently had iodine deficiency. Five of the foals died within 36 hours after birth, one at four days, and one at a week. The diagnosis of iodine deficiency was made on the basis of histologic changes in the thyroid gland. Iodized salt was apparently available on some of the farms, but in at least two instances it appeared that the salt had been buried under hay for several months. On the other farms, the salt blocks may not have been consumed.

On the other hand, excessive iodine intake can be toxic. Drs. Baker and Lindsey (1968) reported that when pregnant mares consumed 48 to 55 mg of iodine (about 50 times the requirement), 5 of 165 foals had congenital goiter. On a farm in which the mares consumed 288 to 432 mg of iodine, 50 percent of the foals had congenital goiter. Some of the affected foals had contractions of the flexor tendons and some died at birth. Excessive iodine may have been transported across the placenta, inhibiting thyroxine synthesis and causing the goiter. Nursing foals can also obtain a substantial amount of iodine from the milk. Kelp contains a high level of iodine; a substantial amount of it was fed to these mares.

Goiter in foals due to excessive iodine intake has been reported by several other authors. In one case the owner was feeding twelve times the manufacturer's recommended level of a supplement containing seaweed. Fortunately in most cases, the foals will recover if iodine is removed from the diet, unless the condition is too far advanced.

MANGANESE

Manganese is required for enzymes that are needed for the formation of cartilage; thus manganese-deficient animals may have shortened and crooked limbs. The bones of the inner ear may not develop properly, causing deafness. Field cases of manganese deficiency have been reported in crippled chickens, characterized by slipped tendon or perosis. The Achilles tendon slips from its condyle or place on the bone because of the malformation of the bone.

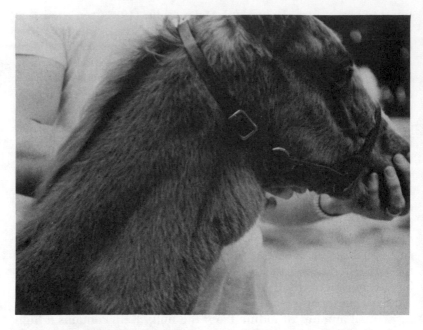

Foal with enlarged thyroid because of excessive iodine intake by the mare. (Courtesy Florida Diagnostic Laboratory, Florida Department of Agriculture)

Manganese deficiency has not been frequently reported but it cannot be discounted. Ranchers in Oklahoma near a zinc smelter found that they had to add high levels of limestone to the soil to combat the high acidity caused by the sulfur dioxide. Alfalfa from the heavily limed fields contained about twice as much calcium and about one third as much manganese as usual, suggesting that the high level of calcium interfered with manganese absorption by the plant. Dr. U. M. Cowgill and co-workers (1980) reported that mares fed the alfalfa during pregnancy gave birth to foals with legs so misshapen and joints so enlarged that they were unable to flex them. The limb bones in some foals were quite shortened. The backbone was abnormally curved. Some foals appeared to be deaf because of abnormal ear development. The skulls were often asymmetric. The animals could not nurse properly and subsequently died.

The manganese requirement of horses is not known, but based on studies of other species, it is estimated to be about 40 ppm. The alfalfa from affected ranches contained levels of manganese as low as 13 ppm.

Further studies at the University of Pennsylvania suggest that the sandy soil in southern New Jersey contains a low level of manganese. Hay and pasture from that area is deficient in manganese and may be the cause of skeletal problems.

Normal blood levels of manganese are low, so it is often difficult to detect manganese deficiency in animals. Manganese appears in several organs, skin, muscle, and bones, but principally in the liver, so liver biopsies indicate nutritional status.

ZINC

Zinc is found in high concentrations in epidermal tissues such as skin and hair, but it also occurs in bone, muscle, blood, and internal organs. Zinc is required for several enzyme systems; its deficiency was first observed in swine in 1955. Pigs fed zinc-deficient diets or diets in which high levels of calcium interfered with zinc utilization, developed parakeratosis. The skin was rough and scaly, and growth was retarded. Subsequent studies with chickens demonstrated that zinc deficiency caused shortened and thickened long bones and poor feathering.

Dr. D. Harrington and co-workers (1973) reported that foals fed zinc-deficient diets showed reduced rate of growth, alopecia (loss of hair), cutaneous lesions on the lower extremities (skin was rough and scaly), reduced tissue and blood zinc levels, and reduced blood alkaline phosphatase activity.

The NRC estimated requirement is 40 ppm, but recent studies indicate that 30 ppm or lower may be adequate.

Zinc deficiency of horses is not commonly reported in field cases, but it may become more of a problem. Zinc toxicity can also be a problem in some situations. Foals fed 90 grams of zinc daily developed enlarged epiphyses followed by stiffness, lameness, and increased tissue zinc levels. Horses grazed on pastures near industrial complexes that have significant zinc emissions may be exposed to levels as high as 90 grams or higher per day.

COPPER

Copper has many functions. It is needed for bone formation, cartilage and elastin formation, utilization of iron, and formation of the pigments in the hair. It is also needed for tyrosinase activity to form the pigment melanin. Studies with other species show that the bones of severely copper-depleted animals have a thin cortex and may break easily. The major blood vessels may rupture because the elastin lacks strengthening cross-links. The animals become anemic because of decreased hemoglobin formation. Coat color will fade: Herford cattle develop a pale yellowish coat, Angus appear dirty grey.

Copper deficiency in horses may not be widespread, but several cases have been reported. Dr. R. B. Becker and co-workers (1957) in Florida noted that foals raised in areas known to be copper-deficient for cattle have swellings at the end of the leg bones. There was loss of angulation in fetlocks and pasterns. Carberry (1978) from New Zealand reported that a foal had painful swellings of the fetlocks consisting of exostoses (bony outgrowths) above and below the joints. The foal was anemic and serum copper level was below normal. Cattle on the same property showed signs of deficiency if not treated regularly with copper.

Drs. Egan and Murrin (1973) of Ireland reported that a Thoroughbred foal was lame and had exostoses above and below the fetlocks, but after copper injections all symptoms disappeared. The foal seemingly had an inherited disability to utilize dietary copper, since none of the other foals on the farm had problems and the diet was adequate.

High levels of molybdenum may interfere with copper utilization. Cymbaluk and co-workers (1981) reported that the addition of 107 ppm of molybdenum to a ration containing 10 ppm copper decreased copper absorption by one third and retention by two thirds. Other trace minerals such as zinc may also interfere with copper utilization.

Copper supplements are often advocated as tonics or for other functions. Colorful myths persist. Some horse owners claim that if you feed your mare from a copper bucket or put pennies in her drinking water she will not come into heat.

Excess copper can be toxic, but more so for sheep than for horses. Smith et al. (1975) reported that feeding diets containing more than 100 times the copper requirement to horses and ponies for 6 months caused an accumulation of copper in the liver of ponies, but did not adversely affect mares or foals. Sheep fed levels of 5 to 10 times the copper requirement become weak, arch their backs because of kidney pain, have bloody urine, and become jaundiced. At necropsy the liver and kidneys show severe degenerative changes. The spleen is enlarged with whole or partial red blood cells; the condition is commonly called "blackberry jam." Cattle can tolerate levels as high as 20 times the copper requirement before toxicity develops.

COBALT

Cobalt is a component of vitamin B12; there is no other role known for it in horses. Bacteria in the digestive tract can synthesize B12 if cobalt is present. Some of the B12 produced by the bacteria apparently can be utilized but the extent of this is unknown.

Toxic Minerals

Several minerals are of interest primarily because of their toxic nature.

FLUORINE

Excess fluorine (over 60 ppm) can cause severe skeletal damage. Drs. Sharpe and Olson (1971) reported that horses with marked fluorosis had bones that were thicker than normal with a chalky white, roughened irregular surface. Fetlock joints were enlarged. Teeth were worn down, making chewing difficult and causing excessive slobbering.

Shupe and Olson (1971) listed four sources of excessive fluorides: (1) forages subjected to airborne contamination in areas near certain industrial operations such as aluminum smelters; (2) high-fluoride water from natural or industrial sources; (3) feed supplements and mineral mixtures containing excessive fluoride; (4) vegetation growing in soils high in fluoride.

In areas where fluorosis is a problem because of contaminated hay or pasture, they suggest you:

1. Increase the grain allowance to decrease hay or pasture intake.
2. Mix hay low in fluorine content with the problem hay to make a ration containing less than 60 ppm fluorine.
3. Feed suspect hay to less valuable mature animals and those not being kept for breeding purposes.
4. Avoid full grazing during late fall and winter, when vegetative growth is slow.
5. Chop hay for animals with damaged teeth. Warm drinking water to help decrease pain.

LEAD

Chronic lead poisoning can cause colic, diarrhea, joint stiffness, labored breathing, roaring because of paralysis of the laryngeal and pharyngeal muscles, weight loss, and death.

Authorities disagree on the toxic level of lead. Aronson (1972) says lead poisoning in horses requires 1.7 mg of lead per kilogram of body weight, but 6 to 7 times that amount produces lead poisoning in cows. 80 ppm of lead in forage could be toxic to horses, but cattle could tolerate 200 ppm or more. However, Dollahite (1978) recently reported that horses are more tolerant of lead than are cattle.

Excessive levels of lead can be obtained by horses grazing on pastures contaminated by emissions from mines or smelters, or by ingestion of lead-based paint. Less likely sources are ingestion of storage batteries or linoleum.

Although lead toxicosis is not a widespread problem, it can be of great concern in certain areas. Numerous horses have been treated for lead toxicosis in the lead/silver belt of northern Idaho. A recent survey by Dr. G. E. Burrows and co-workers (1981) indicated that 9 percent of 118 horses tested had high levels of lead in their blood. The incidence of high blood levels probably could have been higher except many owners were aware of the problem and fed only imported feed. Burrows suggested that increasing the calcium and protein levels of the diet might also help reduce the incidence of toxicosis.

MERCURY

Mercury poisoning has occurred in horses mistakenly given seed grains treated with mercury compounds. The kidneys and central nervous system had degenerative lesions, and the horses became immobilized and depressed. The seed grains were treated with mercury to protect against disease when planted.

ARSENIC

Arsenic has an infamous history. Aristotle in 340 B.C. wrote that *sandarac* (arsenic) killed horses. Many animals have died of accidental poisoning because of ingestion of compounds such as lead arsenite (which was used in dips for ticks and mites) and Paris green (which farmers used in insecticidal sprays). The signs are severe colic, diarrhea, weak pulse and respiration, incoordination, convulsions, and subnormal temperature.

Arsenic has been a favorite of murderers because it is odorless and tasteless. The Marquise de Brinvilliers, during the reign of Louis XIV, killed her father, brother, servants, husband, two lovers, eldest daughter, and several hospital patients with arsenic. But arsenic is difficult to destroy. Modern methods can detect arsenic in body tissue years after death and burial. For example, based on hair analysis, it has been concluded that Napoleon was poisoned with arsenic.

Nevertheless, arsenic has its good side. Small amounts can alleviate selenium toxicity. At one time women took arsenic compounds to improve their complexions.

Many horse trainers have given horses arsenic, thinking that it improves bloom and increases energy and hoof growth; however, no controlled studies have been conducted to verify these thoughts.

Recent studies have indicated that arsenic could be considered an essential mineral: Pigs or chickens fed very low levels of arsenic for three generations had reduced fertility and growth rate. But arsenic deficiency does not seem to be a practical concern.

Vitamins

Vitamins are organic compounds required in small amounts for the normal functioning of the body. They are traditionally divided into two groups: fat-soluble (vitamins A, D, E and K) and water-soluble (including thiamin, riboflavin, niacin, pyridoxine (B6), pantothenic acid, biotin, folic acid, vitamin B12, and vitamin C). Vitamin requirements are listed in Table 5.7 in Chapter 5. The vitamin content of selected feeds is shown in Table 3.4.

VITAMIN A

Vitamin A is required to maintain cellular structures. Deficiencies in animals can cause impaired reproduction, fragile bones, excessive lacrimation (continuous tearing), night blindness, loss of appetite, reduced resistance to infection, elevated spinal fluid pressure, convulsions, rough hair coat, and scaly hooves.

Plants do not contain vitamin A; however, they contain the yellow pigment carotene, which body tissues can convert to vitamin A. The conversion efficiency depends on several factors, but the

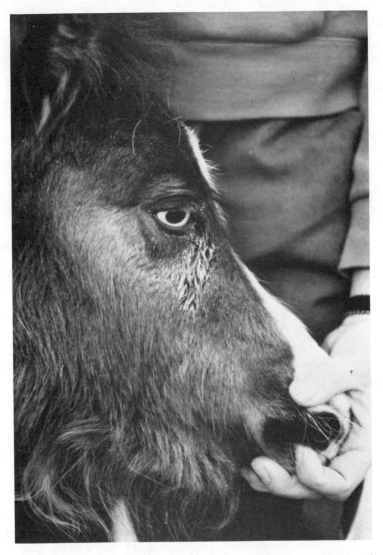

Foal with vitamin A deficiency. Note the excessive lacrimation (tearing) probably caused by changes in cells lining the tear duct. (Photograph by the author)

Table 3.4.   Vitamin Content of Some Feedstuffs[a]

Feedstuff	Carotene (mg/kg)	Riboflavin (mg/kg)	Thiamin (mg/kg)	Vitamin E (mg/kg)
Alfalfa hay, early cut	80	13	3.5	120
mid-bloom	30	10	3.0	80
late bloom	20	9	2.0	60
Alfalfa meal	120	9.5	3.3	120
Barley	—	1.6	4.5	15
Brewers' grains	—	1.2	0.5	25
Brewers' yeast	—	36.0	90	2
Clover hay, ladino	18	13	3.6	65
Clover hay, red	18	15	2.0	55
Corn	—	1.3	2.0	25
Linseed meal	—	2.7	8.0	15
Oats	—	1.5	6.5	16
Rye	—	1.3	3.0	15
Skimmed milk, dehydrated	—	18	3.6	9
Soybean meal	—	3	5.6	3
Timothy hay, early cut	30	15	3.0	60
mid-bloom	15	11	1.5	55
late bloom	12	7	1.0	30
Wheat bran	—	5	6.6	12

[a] 90 percent dry matter basis.

National Research Council has estimated that, for the horse, 1 mg of carotene is equivalent to 400 IU of vitamin A.

Horses do not convert all carotene to vitamin A. Some is stored in the fat; hence horses fed diets containing high amounts of carotene such as leafy alfalfa develop yellow fat. Some of the yellow color in the serum of horses is also due to carotene.

Carotene is rather easily oxidized or destroyed. As much as 80 percent of the carotene in hay can be lost during harvesting and the carotene content of hay stored for six months may be reduced by half. Overheating pelleted feeds can also destroy carotene.

Good pasture is an excellent source of carotene. Properly harvested legume hay has a significant amount of carotene even though much is lost during harvesting. Good quality grass hay can also contain high levels of carotene but not as much as legume hay. Dehydrated alfalfa meal is also an excellent source of carotene.

Corn contains some carotene; the other grains contain little or none. Vitamin A palmitate or other commercially available supplements can be utilized efficiently by horses.

Recent reports from the University of Pennsylvania (Donoghue and Kronfeld, 1981) suggest that the National Research Council recommendations for vitamin A are too low and that horses should be fed two to five times the NRC level, but no more. Studies with ruminants suggest that carotene may have beneficial effects on reproductive performance other than simply as a source of vitamin A. Further studies are needed to determine whether carotene has such effects in horses.

VITAMIN D

Vitamin D is required for the formation of calcium-binding protein, which aids in the absorption of calcium and phosphorus. A deficiency can result in poorly mineralized bones (rickets), swollen joints, stiffness of gait, and reduced serum calcium and phosphorus levels.

Horses normally obtain sufficient vitamin D from sun-cured forages or from exposure to sunlight. The ultraviolet rays convert a compound produced by the body (7-dehydrocholesterol) into vitamin D3. In fact, it is difficult to produce a deficiency in horses fed adequate levels of calcium and phosphorus. Dr. El Shorafa and co-workers (1979) at the University of Florida reported that ponies deprived of vitamin D and sunlight for five months had reduced growth rate and loss of appetite. However, no external signs of rickets were apparent, although bone ash concentration was lower. It was concluded that no dietary vitamin D is needed if horses have sunlight. No estimate of the amount of sunlight needed is available, but in humans some 20 minutes daily is adequate.

Excessive levels of vitamin D can cause calcification of soft tissues such as the blood vessels, heart, lungs, and kidneys; several cases of toxicity have been reported. Certain plants like jessamine can also cause vitamin D toxicity. Krook and co-workers (1975) reported that horses eating jessamine (*Cestrum diurnum*) developed extensive calcium deposits in ligaments, tendons, and blood vessels. All of the horses were lame and some died. The plant was introduced to Florida a few years ago. It spreads rapidly and can also be found in several subtropical areas of Texas and California. *Solanum malocoxylon* and yellow oat-grass (*Trisetum flavescens*) can also cause soft tissue calcification.

VITAMIN E

Vitamin E (tocopherol) is necessary for normal cell structure. Deficiency can cause a wide variety of problems including white muscle disease and anemia. Selenium and vitamin E act synergistically in the prevention of white muscle disease. Supplements of vitamin E are often advocated to improve fertility in mares and stallions.

Alfalfa meal can be an excellent source of vitamin E. Some samples may contain 200 IU/kg; the average is about 135 IU/kg. Pasture is also an excellent source of vitamin E. Good quality hays may contain 50 to 90 IU/kg. Poor quality, weather-damaged hay may contain much lower levels. Grains contain 15 to 30 IU/kg. The NRC estimated requirement is 15 IU/kg.

VITAMIN K

Vitamin K is required for the formation of prothrombin and other clotting factors. A deficiency causes internal hemorrhage. This is rare in horses because the bacteria in the intestinal tract produce significant amounts of vitamin K; however, ingestion of moldy sweet clover containing dicumarol can induce a deficiency.

Recently dicumarol has been used to treat navicular disease. It was theorized that a decreased blood supply to the navicular bone may cause the disease and perhaps anticoagulants such as dicumarol could improve this. But be careful. We found that the clotting time in some horses given dicumarol may unexpectedly increase greatly and the horses may die from internal bleeding.

B COMPLEX

The B complex vitamins are usually supplied in adequate amounts by good quality hay or pasture and by vitamins produced by bacteria of the intestine. However, hardworking horses may require greater amounts than those found in average or poor feeds.

THIAMIN

Thiamin (vitamin B1) appears to be one of the B complex vitamins most likely to be lacking in horse rations. Horses fed poor quality hay can develop signs of thiamin deficiency. We fed ponies hay that had been autoclaved to destroy thiamin. Within a few weeks the blood levels of thiamin and the activity of erythrocyte transketolase (an enzyme which requires thiamin) decreased significantly. Low levels of blood thiamin have been reported in some racing Standardbreds. The "off-feed" that sometimes happens at the racetrack may be effectively treated by giving a thiamin supplement. Cymbaluk et al. (1971) suggested that roaring may be a consequence of thiamin deficiency because roarers have lower blood thiamin levels than non-roarers. Thiamin injections are also often used in the treatment of tieing up. Further research is needed, but racehorses appear to be more susceptible to marginal thiamin deficiency than other horses because they burn up more energy and thiamin is required for energy utilization. Similarly, working horses may require a greater concentration of dietary thiamin than horses at rest.

Thiamin deficiency has been produced in horses fed polished rice and rice straw in Japan. Scientists at the University of California at Davis fed horses a diet containing 68 percent dried beet pulp, 10 percent glucose, 10 percent corn starch, 5 percent casein, 5 percent fishmeal, and 2 percent bone meal, plus a supplement of vitamins A and D. The diet contained 1.1 mg of thiamin per kilogram. Within four months the horses lost their appetite, became nervous and weak, and lacked coordination. One animal died after 19 weeks apparently of cardiac failure; he had an enlarged heart.

The surviving animals were given 30 mg of thiamin daily and recovered.

Thiamin deficiency has been produced in horses eating bracken fern (*Pteridium aquilinium*) or horsetail (*Equistem arrense*). These plants contain compounds antagonistic to thiamin. Both plants are quite common in many parts of the United States, but horses do not normally eat them unless other feedstuffs are lacking.

Brewer's yeast is an excellent source of thiamin. It contains about 30 times as much thiamin as that found in alfalfa hay and 40 times as much thiamin as in corn.

Large doses of thiamin have a stimulating effect on some horses, a tranquilizing effect on others, or no effect at all.

RIBOFLAVIN

Riboflavin (vitamin B2) is also involved in energy metabolism. At one time it was thought that periodic ophthalmia (moon blindness, recurrent uveitis) was caused by riboflavin deficiency. But recent studies suggest that periodic ophthalmia is probably caused by other conditions such as after-effects of leptospirosis or onchocerca cervicalis microfilaria rather than by riboflavin deficiency. Good quality hay or pasture usually contains much higher levels than the estimated requirements.

VITAMIN B12

B12 is a unique vitamin. It is not produced by any plants and it contains cobalt. Mature horses have been fed B12-deficient diets for 11 months without developing deficiency signs. Apparently adequate amounts of B12 can be produced by the bacteria in the cecum and large intestine and absorbed to meet the horse's requirements if cobalt is present. B12 injections are often given to horses, but much of the injected vitamin is excreted in the urine.

Vitamin B12 is required for the normal production of red blood cells. A deficiency can cause anemia, weight loss, reduced performance, and poor hair coat. Humans that cannot absorb B12 because of genetics or other problems develop pernicious anemia.

The vitamin B12 requirement for horses is probably lower than the requirement for cattle because cattle metabolize greater amounts of bacterial products—such as propionic acid—than do horses. Horses have remained in good health while grazing pastures so low in cobalt that ruminants have died there.

FOLIC ACID

Folic acid has many functions, including an important role in the formation of red blood cells. A deficiency of folic acid causes anemia, but red blood cells may be larger than usual.

The folic acid requirements of horses are not known. Horses grazing on good pasture, an excellent source of folic acid, have higher blood folate levels than horses fed hay and grain. But Dr. B. V. Allen of England (1978) suggests that exercise will decrease blood folate levels because significant amounts of folic acid are lost in sweat. Exercise stimulates the formation of red blood cells and increases the need for folic acid within the bone marrow.

NIACIN

Niacin is a component of two coenzymes: nicotinamide adenine dimicleotide (NAD) and nicotinamide adenine dinucleotide (NADP) both of which are important for normal metabolism.

Niacin deficiency is not expected in horses fed the usual feedstuffs. Protein supplements and forages are excellent sources of niacin, and it can be synthesized from the amino acid tryptophan. One study reported that the amount of niacin excreted in the urine can be greater than the amount of niacin consumed. The extra vitamin could come from the conversion of tryptophan to niacin or from bacteria in the intestinal tract.

PANTOTHENIC ACID

Pantothenic acid is a component of coenzyme A, which is an integral part of the metabolism of fat, carbohydrate, and certain amino acids. The requirement for pantothenic acid is not known; based on early studies, however, it has been suggested that 15 ppm in the diet should be adequate (the actual requirement is probably lower). The requirement for baby chicks and growing pigs is about 11 ppm. Good quality hay contains 20 or more ppm, and grains contain about one half of this. Alfalfa pellets may contain 30 ppm or more and pasture can contain 40 or more ppm (dry matter base). A deficiency of pantothenic acid is unusual when horses are fed good quality hay or pasture.

VITAMIN B6

Vitamin B6 (pyridoxine) is concerned with protein metabolism; it is found in forages and grains. B6 deficiency has not been reported in horses, although deficiency in rats causes loss of hair, weight loss, and convulsions.

BIOTIN

Biotin deficiency in pigs causes dermatitis, cracks in the hind feet, and lack of coordination. But deficiencies have not been reported in ruminants or horses, since in these animals there is significant synthesis of biotin by intestinal bacteria.

CHOLINE

Choline is important in fat metabolism and for maintaining cell structure. A choline deficiency may cause fatty liver in chicks and rats, but deficiencies have not been reported in horses.

VITAMIN C

Vitamin C is required for the normal formation of collagen. A deficiency in humans results in scurvy, characterized by swollen, bleeding, and ulcerated gums; loosening of teeth; weak bones; and weakened capillaries, with resulting hemorrhages and delayed wound healing. Fortunately horses, like most species of farm animals, can synthesize vitamin C from glucose. Some species such as humans, guinea pigs, monkeys, bats, certain birds, and fish lack the enzyme L-gluconolactone oxidase, which is needed for the conversion of glucose to vitamin C. These species therefore require vitamin C in the diet.

Vitamin C supplements have been advocated to improve reproductive performance and decrease incidence of epistaxis (nosebleeds) and respiratory diseases, but the claims are unsupported. Similarly, several other vitamins or vitamin-like compounds (bioflavonoids, laetrile, pangamic acid) have been suggested as being of some value to horses—but again, there is little scientific evidence to support such claims, and further research is needed.

VITAMIN SUPPLEMENTS

In most situations if good quality hay or pasture is fed, no vitamin supplements are necessary. Vitamin activity—in particular, vitamin A activity—decreases with the age of the plant, weather damage, and storage. A vitamin supplement is therefore recommended when late cut or weather-damaged hay that is over one year of age is fed.

Numerous supplements are on the market. Some are greatly overpriced and do not provide the amounts and kinds of nutrients needed. Many manufacturers make unrealistic claims.

It is impossible to recommend one supplement because this depends on which nutrients are lacking in the ration. In some cases no one supplement is sufficient and a combination is best. But resist the temptation to oversupplement. "A little bit is good" does not mean that "a whole lot is great."

Several vitamin-mineral supplements were compared in the June 1979 issue of *Practical Horseman*. The cost per day of the manufacturer's recommended daily dose for an average adult horse varied greatly. The cheapest was 12 cents per day and the most expensive was 79 cents per day. The average cost was 35 cents per

Table 3.5.   Twelve Commercial Vitamin-Mineral Supplements
Compared to the NRC Recommended Daily Intakes
for Six-Month Weanlings Weighing 500 lb. (NRC = 100)

Product	Vitamins			Minerals				
	A	D	E	Zn	Mn	Cu	Co	I
A	360	130	740	62	17	60	1,500	600
B	130	200	11	34	14	71	90	360
C	72	26	94	63	14	110	3,930	—
D	680	1,250	23	—	35	9	200	440
E	540	580	74	11	75	22	750	1,500
F	140	270	46	17	8	63	620	—
G	580	250	300	13	—	4	110	1,000
H	1,090	1,670	930	75	100	560	6,000	—
I	200	330	23	10	9	12	820	1,540
J	200	120	3	2	34	5	91	270
K	500	620	87	7	14	60	1,420	62
L	760	360	93	—	2	4	170	340

Adapted from Donoghue and Kronfeld, 1981.
Nutrient analyses, daily doses, and daily costs of these products were tabulated in *Practical Horseman*, June 1979, pp. 74–75.
Excesses of cobalt, iodine, or Vitamin D indicate that eight of the products should be used with caution. The remaining four are not strong in manganese or vitamin E, and none have selenium.

day or $127.75 per year. Donoghue and Kronfeld (1980) compared the nutrient intakes provided by the manufacturers' recommended dose of several supplements with the National Research Council's recommended level for a 6-month old weanling weighing 500 pounds (Table 3.5). Remember, the values are only for the supplement and do not include the nutrients in the hay and grain.

None of the supplements provided a significant amount of calcium. The most expensive supplement supplied amounts of vitamins A, D, E, riboflavin, thiamin, and trace minerals at levels several times that recommended by the National Research Council. There is no evidence to suggest that the higher levels are beneficial; excessive levels of some nutrients can be toxic. The cheapest supplement provided 100 percent of the requirement for vitamin A but much lower percentages of most of the other nutrients; it could have been an effective supplement if the primary concern was vitamin A.

Remember, the nutrient content of the supplement should be determined. Supplements should not be added without a reason, and sometimes a combination of supplements may complement the ration better than one. But don't overdo the manufacturer's recommendation.

Compare prices in relation to the nutrients provided. Some of the materials compared by *Practical Horseman* magazine were much more costly than others, but they did not provide significantly more nutrients. Buy from a high-volume dealer with a rapid merchandise turnover to ensure a fresh product. Vitamin destruction takes place during storage.

Be wary of manufacturers that make outrageous claims for their products.

WATER

Water comprises about 70 percent of the lean adult body; a 1,100-lb horse contains about 90 gallons. Water is important in digestion, body temperature regulation, and as a solvent. It lubricates joints, acts as a cushion for the central nervous system, transports sound, is required for sight, and has many other functions.

There is an old saying that there is no one most important essential nutrient. However, water is one of the most critical nutrients because an animal can live longer without food than it can live without water. After two days without water animals refuse to eat and may show signs of colic. After three days the animals can be very uncomfortable and restless. Experiments conducted in Paris in the summer of 1882 indicated that horses could survive for 5 or 6 days without water but could live 20 to 25 days without food if water was provided. The length of time before water deprivation causes death depends on several factors such as environmental temperature and activity of the horse.

Dehydration can be caused by inadequate water supplies, excessive water loss through diarrhea, sweating, or an inability to drink because of choke, tetanus, or a fractured jaw. The extent of dehydration can be estimated by several methods; clinical examination is useful. First, a dehydrated horse will have loss of skin elasticity. Grasp a pinch of skin on a normal horse and it will immediatley return when released, but a pinch on a mildly dehydrated horse may stand for 2 to 5 seconds and between 5 to 10 seconds or more in a severely dehydrated horse.

In addition, a dehydrated horse will have sunken eyes and a "tucked up" appearance. Capillary refilling time is increased. The hemoglobulin level and packed-cell volume will increase, but excited horses can also have increased values because of the contraction of the spleen releasing red blood cells. The osmality and total plasma protein level will also be increased in a dehydrated horse.

The amount of water required by a horse depends on the water in the feed (grass contains 60 to 75 percent water, hay only 10 percent), environmental temperature, body temperature, activity, and amount eaten. Under most conditions there is a very high correlation between water intake and feed intake. A horse may be expected to drink 1 to 2 quarts of water per pound of feed.

A 1,100-pound horse may drink 6 to 10 gallons of water daily. A hard-working horse may require much more. A lactating mare producing 30 lbs of milk daily would require an additional 5 gallons to that needed for maintenance. But the best policy is to let the horse drink all it wants, provided it is not hot.

We tested preference for temperature of water in six mature pony mares after their morning feeding of hay. The ponies were given a choice of water at 35°F, 50°F, 68°F, or 86°F in an arrangement of two-choice preference tests. 35° was compared to 50°, 68°, or 86°; 50° was compared to 68° or 86°; and 68° to 86°. The temperature in the barn at the time of the tests was usually 40° to 50°F.

Five of the six ponies usually drank more of the colder water in each comparison. The average intake of the colder water for the five ponies averaged 60 percent greater than for the warmer water. One pony consumed equal amounts of the colder and warmer water.

Water weighs 8 pounds per gallon, so it's heavy to haul around. Automatic waterers that are kept clean and are heated to prevent freezing in the winter are a great labor-saving device, but they should be constructed so that the horses do not hit their legs.

Donkeys can tolerate lack of water better than horses and will continue for several days without water. During periods of water deprivation donkeys have increased body temperature and they produce drier feces, but they have a remarkable ability to recover. One study in Africa showed that donkeys dehydrated to only 85 percent of their original weight could drink enough water to regain all this lost weight within 3 to 5 minutes.

Donkeys effectively retain the capacity to secrete saliva even when severely dehydrated, enabling them to continue eating. In the desert they can graze for 24 hours before returning to a water source. In one study, donkeys chose the hay first after being without hay or water for 24 hours. However, when they were more dehydrated they chose the water first. Ideally, of course, donkeys should be provided water freely.

Horses and donkeys have been known to exist on snow as the only source of water for limited periods, but preferably the animals should have other sources of water.

A group of horses were stranded in a hunting camp in Alaska because of early snow. The animals were fed by feed dropped from an airplane, but no water other than snow was available from October 2 to the middle of February, when the animals were rescued. Although some of the horses died, the cause of death was because of the severe weather and lack of food, not from lack of water. There was an accumulation of 40 inches of snow and the average temperature was 0°F.

In controlled studies by Drs. R. A. Dieterich and E. A. Holleman (1973), feed intake was similar for horses given only snow or only water for several weeks and the weight changes were not different, indicating that the caloric cost of consuming snow was not great. Biochemical blood values were also similar and were within normal ranges. The authors concluded that horses have remarkable physical stamina.

Quantity is important but quality must also be checked. Stagnant water in ponds can harbor diseases. Some water contains potentially toxic levels of minerals such as fluoride and selenium. The

Table 3.6.   Recommended Upper Limits of Some Toxic Substances in
Drinking Water

Substance	Safe Upper Limit (mg/liter)	Substance	Safe Upper Limit (mg/liter)
Arsenic	0.2	Mercury	0.01
Cadmium	0.05	Nickel	1.0
Chromium	1.0	Nitrate	100
Cobalt	1.0	Nitrite	10
Copper	0.5	Vanadium	0.1
Fluoride	2.0	Zinc	25.0
Lead	0.1		

National Research Council summarized the recommended upper limits of some toxic substances, shown in Table 3.6.

Excess total salt in water is a problem in many parts of this country, particularly in the West. For example, 74 percent of the surface water and 8 percent of the ground water samples tested in South Dakota had unacceptable levels of salt. A measure of conductivity gives an estimate of total dissolved solids (TDS) or salt. Water with TDS of 1,000 ppm or less should cause no problems. If animals are changed suddenly to water containing 1,000 to 5,000 ppm TDS, diarrhea may develop at first, but the animals should adapt. Water with 5,000 to 7,000 ppm TDS may not greatly influence performance, but water over 7,000 ppm TDS can impede growth rate and performance.

References

Allen, B.V. Serum folate levels in horses, with particular reference to the English Thoroughbred. *Vet. Rec.* 103:257, 1978.

Arnbjerg, J. Poisoning in animals due to oral application of iron. *Nord. Vet.-Med.* 33:71, 1981.

Aronson, A.L. Lead poisoning of cattle and horses following long-term exposure to lead. *Am. J. Vet. Res.* 33:627, 1972.

Baker, H.J. and J.R. Lindsey. Equine goiter due to excess iodine. *JAVMA* 153:1618, 1968.

Becker, R.B. et al. Minerals for dairy and beef cattle. *Univ. Fla. Ext. Bull.* 513R, 1957.

Bitting, A.W. Big-Head. *Florida. Ag. Exper. Stat. Bull.* 26, 1894.

Buntain, B.L. and J.R. Coffman. Polyuria and polydypsia in a horse induced by psychogenic salt consumption. *Eq. Vet. J.* 13:266, 1981.

Burrows, G.E. A survey of blood lead concentrations in horses in the North Idaho lead/silver belt area. *Vet. Human Toxicology* 23:328, 1981.

Carbery, J.T. Osteodysgenesis in a foal associated with copper deficiency. *New Zealand Vet. J.* 26:279, 1978.

Carrol, F.D., H. Goss, and C.E. Howell. The synthesis of B-vitamins in the horse. 8:290, 1949.

Colles, C.M. The use of warfarin in the treatment of navicular disease. *Equine Vet. J.* 11:187, 1979.

Corke, H.J. An outbreak of sulphur poisoning in horses. *Vet. Rec.* 169:272, 1981.

Cowgill, U.M. et al. Smelter smoke snydrome in farm animals and manganese deficiency in northern Oklahoma. *Environ. Poll.* 22:259, 1980.

Cymbaluk, N.F., P.B. Fritz, and F.M. Loew. Thiamin measurements in horses with laryngeal hemiplegia. *Vet. Rec.* 101:97. 1977.

Cymbaluk, N.K. et al. Infleunce of dietary molybdenum on copper metabolism in ponies. *J. Nutr.* 111:96, 1981.

Dieterich, R.A. and D.E. Holleman. Hematology, biochemistry and physiology of environmentally stressed horses. *Can. J. Zoo.* 51:867, 1973.

Doige, C.E. and B.G. McLaughlin. Hyperplastic goitre in newborn foals in Western Canada. *Can. Vet. J.* 22:42, 1981.

Dollahite, J.W. et al. Chronic lead poisoning in horses. *Amer. J. Vet. Res.* 39:961, 1978.

Donoghue, S. and D.S. Kronfeld. Nutritionally related bone diseases. *Proc. Amer. Assoc. Equine Pract.* Anaheim, Calif, p. 65, 1980.

Donoghue, S. and D. Kronfeld. Vitamin-mineral supplements for horses. *Comp. Cont. Educ.* 2:5121, 1980.

Donoghue, S. and D. Kronfeld. Vitamin A nutrition of the equine. *J. Nutr.* 111:365, 1981.

Egan, D.A. and M.P. Murrin. Copper-responsive osteodysgenesis in a Thoroughbred foal. *Irish Vet. J.* 27:61, 1973.

El Shorafa, W.M., J.P. Feaster, E.A. Ott, and R.L. Asquith. Effect of vitamin D and sunlight on growth and bone development of young ponies. *J. Anim. Sci.* 48:882, 1979.

Elinder, C. et al. Water hardness in relation to cadmium accumulation and microscopic signs of cardiovascular disease in horses. *Arch. Environ. Health* 35:81, 1980

Harrington, D.D. et al. Clinical and pathological findings in horses fed zinc deficient diets. *Proc. Third Equine Nutr. Phys. Symp.* Gainesville, Fla., p. 51, 1973.

Harrington, D.D. Pathologic features of magnesium deficiency in young horses fed purified rations. *Am. J. Vet. Res.* 35:503, 1974.

Henneke, D.R., G.D. Potter, and J.L. Kreider. A condition score relationship to body fat content of mares during gestation and lactation. *Seventh Equine Nutr. Phys. Symp.* Warrenton, Va., p. 105, 1981.

Hudson, R.S. Salt for horses. *Mich. State Coll. Quart. Bull.* 8:103, 1926.

Jordan, R.M. Sulfur levels for horses. Presented at Seventh Equine Nutr. Phys. Symp. Warrenton, Va., 1981.

Kinter, J.H. and R.L. Holt. Equine osteomalacia. *Philippine J. Sci.* 49:1, 1932.

Krook, L. et al. Cestrum diurnum poisoning. *Cornell Vet.* 65:557, 1975.

Maylin, G.A. et al. Selenium and Vitamin E in horses. *Cornell Vet.* 70:272, 1980.

Shupe, J.L. and A.E. Olson. Clinical aspects of fluorosis in horses. *JAVMA* 158:167, 1971.

Slade, L. Blood and tissue chemistry relationships to nutritional status of horses and other animals. In: *Stud Managers Handbook* no. 16, p. 193. Agriservices Foundation, Clovis, Calif., 1980.

Smith, J.D. et al. Tolerance of ponies to high levels of dietary copper. *J. Anim. Sci.* 41:1645, 1975.

Stowe, H.D. Effects of potassium in a purified diet. *J. Nutr.* 101:629, 1971.

Trelease, S.F. and O.A. Beath. *Selenium*. N.Y.: Academic Press, 1949.

Van Soest, P.J. *Nutritional Ecology of the Ruminant*. Corvallis, Oregon: O & B Books Inc., 1982.

Wilcox, E.V. Selenium versus General Custer. *Agr. Hist.* 18:105, 1944.

Chapter Four

Feeds

Knowledge of the characteristics of feeds is essential to keep horses healthy and allow them to perform to their genetic potential.

Certain feeds have long been favorites of horsemen, and tradition has played an important part in this. But the horse can be successfully fed a variety of feedstuffs; flexibility in selection may even reduce the cost of feeding.

The good old days of feeding a horse for 90 dollars a year have long since gone, and in recent years the costs have skyrocketed—even if you grow your own feed. Dr. D. P. Snyder of the Department of Agricultural Economics at Cornell University reported that the cost of growing hay in New York was 17 percent more in 1980 than in 1979; the cost of growing corn increased by 21 percent in the same period. The cost of raising or buying feed will continue to increase, so the more you know about feeds, the more money you'll save.

The nutrient composition of many feedstuffs is given in Table 4.1. The vitamin content of selected feedstuffs is shown in Table 3.4.

Forages

For those whose pleasure it is to rear horses it is of the utmost importance to provide a painstaking overseer and plenty of fodder.—Columella, A.D. 50

HAY

There are two basic kinds of hay: legume and grass. A legume has bacteria in its root nodules that can utilize nitrogen from the air and produce higher levels of protein than levels found in grasses (Table 4.2). When the proper bacteria are present legumes build up the nitrogen content of the soil. Legumes also have a much higher calcium content than grasses and possibly a higher phosphorus content, depending on soil phosphorus levels.

In a report from the New York Dairy Improvement Corporation, samples of legume hay contained slightly more potassium and iron than grass hay, but the grass hay contained slightly more manganese. The magnesium, zinc, and copper contents of legume and grass hays were similar.

The vitamin A activity and B vitamin content of good legume hay is usually much higher than that of grass hay. Legumes will usually produce more nutrients per acre than grass hay. When seeded with grasses, legumes often increase the yield of grasses because of the increased nitrogen supply in the soil provided by the legume bacteria.

What kind of hay is best for the horse?: Hay containing the most nutrients per dollar—provided it is readily consumed and does not contain any harmful ingredients. The feeding of good quality hay makes the job of balancing the horse's rations much easier. Good hay provides energy, protein, vitamins, and minerals. The quality of the hay can be judged by several factors, the most important being the stage of maturity at the time of cutting.

As the plant matures the ratio of stem to leaf increases. The stem contains relatively more structural components (fiber) that are not efficiently digested. The leaf of the legume does not greatly change in composition as it ages, but as the leaf of the grass plant matures the ratio of fiber to non-fibrous fractions increases. Digestibility decreases because the concentration of the more digestible nutrients is decreased.

One study showed that the protein content of boot stage timothy was 13 percent compared to 5 percent in mature timothy. Digestibility of protein in the mature timothy was only half of that

Table 4.1.   Estimated Composition of Various Feeds
PART 1: FEEDS WITH A DRY MATTER CONTENT OF 89 TO 92%
—VALUES ON AS-FED BASIS

Feed	DE	Crude Protein	Lysine	AD Fiber	Calcium
	(Mcal/lb)	(%)	(%)	(%)	(%)
ALFALFA					
Hay, early bloom	1.00	15.5	0.85	34	1.55
Hay, mid-bloom	0.92	14.5	0.80	36	1.35
Hay, full-bloom	0.86	13.0	0.70	38	1.20
Meal, dehy. 15% protein	1.00	15.0	0.85	35	1.30
Meal, dehy. 17% protein	1.02	17.0	0.95	32	1.40
BAHIAGRASS					
Hay, mature	0.84	5.2	NA	NA	0.40
BAKERY PRODUCT, DRIED	1.75	10.0	0.30	NA	0.10
BARLEY					
Grain	1.50	12.0	0.45	6	0.04
Grain, Pacific Coast	1.48	10.0	0.38	8	0.04
Hay	0.75	8.0	NA	NA	0.18
Straw	0.60	3.5	NA	54	0.20
BEET PULP	1.20	7.2	0.60	28	0.65
BERMUDAGRASS, COASTAL					
Immature	0.95	12.5	0.60	34	0.40
Mature	0.80	7.0	0.32	42	0.28
BLOODMEAL, DRIED	1.20	85.0	5.3	NA	0.28
BREWERS' GRAINS	1.25	25.0	0.85	21	0.27
BROME GRASS, HAY					
Hay, immature	0.95	11.0	NA	35	0.45
mature	0.80	6.0	NA	44	0.28
CANARY GRASS REED					
Hay, immature	0.95	13.0	NA	NA	0.40
mature	0.80	6.5	NA	NA	0.30

Table 4.1.   PART 1: (Continued)

Copper (mg/kg)	Iron (mg/kg)	Magnesium (%)	Manganese (mg/kg)	Phosphorus (%)	Potassium (%)	Zinc (mg/kg)
13	180	0.27	30	0.24	2.30	16
12	165	0.26	28	0.23	1.80	16
11	160	0.26	27	0.22	1.65	16
12	180	0.27	29	0.23	2.10	16
12	180	0.27	29	0.25	2.30	17
NA	NA	0.23	NA	0.20	1.30	NA
NA	NA	NA	35	0.35	0.10	15
8	70	0.13	17	0.35	0.45	15
8	70	0.12	17	0.35	0.50	15
4	270	0.17	30	0.27	1.30	NA
9	270	0.13	25	0.10	1.25	NA
12	300	0.27	35	0.09	0.18	9
NA	NA	0.18	NA	0.25	1.70	18
NA	NA	0.16	NA	0.20	1.50	18
10	380	0.22	5	0.22	0.90	NA
22	245	0.15	36	0.54	0.08	20
8	100	.18	NA	0.30	2.20	NA
7	90	.18	NA	0.20	1.80	NA
9	140	0.28	NA	0.26	2.50	NA
8	135	.24	NA	0.22	1.75	NA

Table 4.1. PART 1: (Continued)

Feed	DE (Mcal/lb)	Crude Protein (%)	Lysine (%)	AD Fiber (%)	Calcium (%)
Citrus pulp	1.25	6.3	NA	23	1.80
Clover, Alsike mid-bloom	0.90	13.5	NA	36	1.30
Clover, Crimson, mid-bloom	0.90	13.5	NA	37	1.28
Clover, Ladino, mid-bloom	0.90	13.5	NA	36	1.32
Clover, Red, mid-bloom	0.90	13.5	NA	36	1.35
late bloom	0.80	11.0	NA	39	1.05
Corn, grain	1.60	9	0.27	2.2	0.02
Corn cobs	0.55	2.6	NA	35	0.10
Corn and cob meal	1.35	8	NA	9	0.03
Cottonseed hulls	0.60	3.6	NA	63	0.13
Cottonseed meal	1.40	41	1.65	NA	0.17
Distillers grains, corn	1.55	27	0.9	NA	0.08
HOMINY FEED	1.65	11.5	0.45	2	0.05
LESPEDEZA, mature	0.80	14.0	—	NA	1.10
LINSEED MEAL					
expeller	1.50	32	1.1	NA	0.4
solvent	1.50	33	1.1	NA	0.35
MILK, SKIM DRIED	1.65	32	2.4	—	1.30
OATS					
grain	1.35	12	0.4	NA	0.09
grain, Pacific Coast	1.32	9	0.35	NA	0.10
straw	0.55	4	NA	48	0.25
hulls	0.50	5	NA	NA	0.16
ORCHARDGRASS					
hay, immature	0.95	12	NA	NA	0.40
mature	0.80	8	NA	NA	0.30

Table 4.1.   PART 1: (Continued)

Copper	Iron	Magnesium	Manganese	Phosphorus	Potassium	Zinc
(mg/kg)	(mg/kg)	(%)	(mg/kg)	(%)	(%)	(mg/kg)
5	150	0.14	6	0.12	0.70	14
6	200	0.28	50	0.22	2.20	16
NA	170	0.25	NA	0.21	2.10	15
8	250	0.24	NA	0.23	2.30	17
10	270	0.30	50	0.23	2.30	15
9	250	0.25	40	0.21	2.20	14
3	27	0.02	5	0.30	0.30	18
6	200	0.06	5	0.04	0.80	NA
4	45	0.03	5	0.25	0.40	NA
13	135	0.12	9	0.07	0.81	14
15	90	0.36	18	0.90	1.20	55
40	180	0.05	18	0.40	0.15	NA
14	60	0.24	15	0.50	0.70	NA
NA	300	0.23	165	0.23	1.00	NA
26	200	0.6	39	0.8	1.25	NA
25	300	0.6	38	0.75	1.30	NA
1	10	0.13	2	1.0	1.50	60
6	70	0.15	36	0.35	0.40	30
5	80	0.15	36	0.35	0.40	NA
9	180	0.15	32	0.06	2.00	NA
5	10	0.08	8	0.19	0.60	NA
13	100	0.22	38	0.29	3.00	15
12	100	0.20	35	0.25	2.70	15

Table 4.1.   PART 1: (Continued)

Feed	DE	Crude Protein	Lysine	AD Fiber	Calcium
	(Mcal/lb)	(%)	(%)	(%)	(%)
PANGOLAGRASS					
immature	9.90	12	NA	NA	0.35
mature	0.78	8	NA	NA	0.30
PEANUT MEAL	1.50	45	1.6	NA	0.15
RAPESEED MEAL					
expeller	1.40	36	1.8	NA	0.70
solvent	1.38	36	2.0	NA	0.65
RYE, grain	1.45	12.5	0.4	NA	0.06
SORGHUM, grain	1.55	11.0	0.25	NA	0.04
SOYBEANS					
whole	1.85	38	2.4	NA	0.30
meal, expeller	1.55	44	2.7	NA	0.25
meal, solvent	1.55	44	2.9	NA	0.20
meal, dehulled	1.58	49	3.2	NA	0.25
SPELT, grain	1.25	11.0	0.27	NA	0.03
SUNFLOWER					
meal	1.00	24.0	0.95	NA	0.30
meal, dehulled	1.30	42.0	1.7	NA	0.40
TIMOTHY					
early head	1.00	11	0.55	37	0.35
head	0.85	8	0.40	45	0.25
over ripe	0.75	5	0.18	54	0.18
TREFOIL, BIRDSFOOT					
early bloom	1.0	15.0	0.75	NA	1.70
late bloom	0.83	12.5	0.62	NA	1.30
TRITICALE, grain	1.52	14.0	0.60	NA	0.05
WHEAT					
bran	1.25	15.0	0.60	12	0.10
grain, hard red	1.57	13.0	0.40	4	0.04
grain, soft red	1.57	12.0	0.50	4	0.04
grain, soft white	1.57	10.5	0.30	4	0.04
hay	0.80	7.5	NA	36	0.12
straw	0.60	3.5	NA	46	0.18
YEAST					
brewers	1.40	44.0	3.0	NA	0.40

Table 4.1.   PART 1: (Continued)

Copper	Iron	Magnesium	Manganese	Phosphorus	Potassium	Zinc
(mg/kg)	(mg/kg)	(%)	(mg/kg)	(%)	(%)	(mg/kg)
NA	NA	0.14	NA	0.25	NA	NA
NA	NA	0.12	NA	0.20	NA	NA
NA	NA	0.30	30	0.55	1.0	NA
7	180	0.50	45	1.0	NA	66
7	180	0.50	45	0.9	NA	66
7	60	0.13	50	0.32	0.47	32
15	45	0.18	16	0.30	0.36	14
14	70	0.30	27	0.6	1.60	14
36	150	0.27	27	0.65	1.80	50
36	120	0.27	27	0.65	1.80	50
36	120	0.27	27	0.65	1.85	50
NA	NA	NA	NA	NA	NA	NA
NA	NA	NA	NA	0.9	NA	NA
40	35	0.75	24	1.0	1.0	NA
6	180	0.15	40	0.25	1.80	NA
5	180	0.14	40	0.23	.70	NA
4	150	0.10	30	0.14	1.20	NA
8	200	0.30	NA	0.23	1.60	NA
8	200	0.30	NA	0.20	1.40	NA
NA	NA	NA	NA	0.30	NA	NA
12	170	0.55	115	1.25	1.45	110
5	35	0.15	36	0.45	0.42	40
6	35	0.10	36	0.40	0.38	45
7	35	0.10	36	0.40	0.38	30
NA	180	0.10	36	0.18	0.90	NA
3	180	0.10	36	0.07	1.00	NA
32	90	0.21	5	1.35	1.70	36

PART 2: FEEDS CONTAINING A SIGNIFICANT AMOUNT OF MOISTURE—VALUES ON TOP LINE ARE ON AS-FED BASIS. VALUES IN BOLD FACE ARE ON 90% DRY MATTER BASIS TO ALLOW COMPARISON TO VALUES IN PART 1.

Feed	DE (mcal/ kg)	Crude Protein (%)	Lysine (%)	AD Fiber (%)	Calcium (%)
ALFALFA					
Pasture-prebloom (21)[b]	0.24	4.5	0.22	NA	0.48
90%	**1.05**	**19.0**	**0.90**	NA	**2.00**
Pasture-full bloom (25)[b]	0.26	4.0	0.16	NA	0.38
90%	**0.93**	**14.7**	**0.60**	NA	**1.38**
Silage (28)[b]	0.28	5.0	0.24	NA	0.40
90%	**0.95**	**16.0**	**0.75**	NA	**1.30**
BERMUDAGRASS					
Pasture (39)[b]	0.38	3.5	NA	NA	0.19
90%	**0.90**	**8.0**	NA	NA	**0.44**
BLUEGRASS					
Pasture early (31)[b]	0.35	5.3	NA	NA	0.17
90%	**1.00**	**15.3**	NA	NA	**0.50**
Silage (35)[b]	0.39	5.0	NA	NA	0.25
90%	**1.00**	**13.5**	NA	NA	**0.60**
CLOVER, LADINO					
Pasture (22)[b]	0.24	4.4	NA	NA	0.36
90%	**1.00**	**18.0**	NA	NA	**1.80**
CLOVER, RED					
Pasture (20)[b]	0.22	4.0	NA	NA	0.40
90%	**1.00**	**18.0**	NA	NA	**1.80**
CORN SILAGE					
Milk stage (24)[b]	0.33	1.9	NA	NA	0.07
	1.25	**7.0**	NA	NA	**0.25**
Dough stage (29)[b]	0.42	2.6	NA	NA	0.09
	1.30	**8.0**	NA	NA	**0.30**
MOLASSES					
Beet (79)[b]	1.15	7.6	NA	—	0.10
90%	**1.30**	**8.6**	NA	—	**0.11**
Cane (74)[b]	1.05	2.9	NA	—	0.80
90%	**1.30**	**3.5**	NA	—	**0.98**
Citrus (68)[b]	0.98	5.7	NA	—	1.20
90%	**1.30**	**7.6**	NA	—	**1.60**

Table 4.1. PART 2: (Continued)

Copper (mg/kg)	Iron (mg/kg)	Magnesium (%)	Manganese (mg/kg)	Phosphorus (%)	Potassium (%)	Zinc (mg/kg)
2	41	.05	6	.07	0.50	4
9	180	.23	25	.32	2.10	16
2	41	.04	5	.07	0.54	4
9	180	.21	24	.32	1.94	14
3	45	0.08	8	.08	0.60	4
9	180	.30	28	.32	1.94	14
NA	NA	.07	NA	.10	NA	NA
NA	NA	.17	NA	.24	NA	NA
3	NA	.06	24	0.12	0.70	NA
9	NA	.18	72	0.36	2.00	NA
3	NA	.06	24	0.14	0.70	NA
9	NA	.18	72	0.36	1.90	NA
?	60	0.7	24	0.07	0.60	5
9	250	0.30	100	0.28	2.50	20
NA	60	0.10	NA	0.08	0.50	NA
NA	270	0.45	NA	0.36	2.25	NA
38	15	0.07	8	0.06	0.30	5
14	58	0.25	30	0.21	1.20	18
14	19	0.10	10	0.08	0.40	5
13	60	0.32	35	0.25	1.25	15
18	100	0.23	5	0.20	5.00	NA
20	114	0.26	6	0.23	6.00	NA
60	200	0.35	42	0.08	2.00	NA
74	245	0.44	51	0.10	2.40	NA
74	340	0.14	26	0.12	0.10	NA
100	455	0.19	35	0.16	0.13	NA

Table 4.1. PART 2: (Continued)

Feed	DE (mcal/ kg)	Crude Protein (%)	Lysine (%)	AD Fiber (%)	Calcium (%)
Corn starch (73)ᵇ	1.05	0.5	NA	—	0.10
90%	1.30	0.6	NA	—	0.12
Wood (66)ᵇ	0.95	0.7	NA	—	0.50
90%	1.30	0.95	NA	—	0.68
ORCHARD GRASS					
Pasture (19)ᵇ	0.20	3.4	NA	7	0.11
90%	0.95	16.0	NA	31	0.50
TIMOTHY					
Pasture, mid-bloom (30)ᵇ	0.30	2.8	NA	12	0.15
90%	0.90	8.5	NA	37	0.45

ᵇ Dry matter content on as-fed basis.
NA = Analysis not available.
— = No measurable amount.

in the boot stage. Examples of the effect of maturity on composition are shown in Tables 4.3 and 4.4. Not only do digestibility and nutrient concentration decrease as the plant matures, but the late-cut, mature plants are often unpalatable and are rejected or wasted. Suggestions on when to harvest crops for hay are shown in Table 4.5.

When selecting hay for purchase, have it analysed by a laboratory to measure nutritional value. You can contact your county agent or farm advisor for the name of the forage testing laboratory nearest you. Remember, the analysis is only as good as the sample; simply grabbing a handful of hay from one bale is unlikely to give representative values. Dr. D. A. Rohweder of the University of Wisconsin suggested the following guidelines for sampling:

1. Test each forage lot and each field harvested separately.
2. Take at least 12 widely separated samples from each lot.

Table 4.1. PART 2: (Continued)

Copper (mg/kg)	Iron (mg/kg)	Magnesium (%)	Manganese (mg/kg)	Phosphorus (%)	Potassium (%)	Zinc (mg/kg)
—	—	—	—	0.60	0.20	NA
—	—	—	—	0.75	0.25	NA
—	—	0.07	13	0.05	0.04	NA
—	—	0.10	18	0.07	0.05	NA
1	32	0.04	8	0.09	0.60	3
6	150	0.17	36	0.45	2.70	15
2	60	0.05	NA	0.07	0.72	NA
5	180	0.14	NA	0.22	2.15	NA

3. Always sample with a bale corer to reduce error. A Penn State Forage Sampler is good; it works best with a $1/2$-inch electric drill but may be used with a $3/8$-inch drill or a hand drill. Remove hay from the sampler by disengaging the barrel from the chuck adapter. It is impossible to get a representative sample by using bale slices, etc.
4. Insert the sampler to full depth in loose bales and 12 to 15 inches into the end of each tight bale sample to insure getting both stems and leaves.
5. Use the corer for loose and chopped hay too.
6. Mix the 12 cores in a clean pail and place in a tight, clean plastic or paper bag or other container.

What information can be obtained from forage testing laboratory reports? The number of procedures used will vary among laboratories, but values for dry matter, crude protein (nitrogen times 6.25), and acid detergent fiber are perhaps the most common.

Table 4.2. Composition of Legume, Mixed Mainly Legume, Mixed Mainly Grass and Grass Hays[a]

Item	Legume (%)	Mainly Legume (%)	Mainly Grass (%)	Grass (%)
Dry matter	88.7 (86.2–91.1)[b]	88.8 (86.3–91.2)	89.5 (87.6–91.3)	89.6 (87.6–91.5)
Crude protein	17.4 (13.7–21.0)	15.1 (11.4–18.7)	10.5 (7.2–13.7)	9.5 (6.3–12.6)
ADF	37.9 (31.5–44.2)	38.1 (33.7–42.4)	40.4 (37.2–43.5)	40.9 (37.5–44.2)
Calcium	1.12 (.78–1.44)	.99 (.64–1.34)	.65 (.39–90)	.52 (.30–.75)
Phosphorus	.28 (.20–.35)	.26 (.21–.36)	.22 (.17–26)	.21 (.15–.26)

[a] From the New York Dairy Herd Improvement Corporation Forage Testing Laboratory for the period September 1, 1971 through January 31, 1980. Values expressed on a dry matter basis. Samples are only those from New York State.
[b] Value in parentheses is the normal range (± one deviation from the mean). Approximately 67% of all samples would be expected to be in the normal range.

Drill and core sampler. (Photograph by the author)

Estimated TDN, digestible energy, or even net energy values might also be reported—but be careful because they usually apply to ruminants. Values for roughages—particularly those of poor quality or high fiber content—would overestimate the value for horses. Nevertheless, the estimated value provides a basis for comparison to other roughages. Determination of minerals such as calcium, phosphorus, potassium, iron, manganese, magnesium, zinc, and copper may not be expensive. Selenium and iodine determi-

Table 4.3.   Effect of Harvest Date on Chemical Composition
and Digestibility of First Cutting of Timothy

Date of Harvest	Crude Protein (%)	NDF (%)	ADF (%)	Cellulose (%)	Lignin (%)	Digestibility of Organic Matter (%)
June 17	22	49	27	20	4	80
June 23	16	50	27	21	4	80
July 1	13	61	33	27	5	74
July 7	11	64	36	30	5	71
July 13	10	64	37	30	6	68
July 20	8	64	38	31	6	63

[a] Adapted from data presented by Lindgren et al., *Swedish J. Agric. Res.* 10:3–10, 1980.
[b] All composition expressed on dry matter basis.
[c] Neutral detergent fiber.
[d] Acid detergent fiber.

nations are more difficult and more expensive. Many of the vitamin assays are also expensive.

The forage testing laboratory may not only provide values but also indicate the normal ranges for your area. The values in Table 4.2 show the average and range for samples from New York State.

Another problem with forage analysis is that even when representative samples of the hay offered are taken, they may not accurately represent the feed consumed. In some circumstances (such as when hay is the major or only source of nutrients) the animals may select the leaves and waste the stem. In such a case, the sample would underestimate the concentration of energy, protein, and minerals in the consumed feed because leaves contain a higher percentage of these nutrients than do the stems. If the ration is primarily grain, the animals may eat relatively more stems from the hay, perhaps because they want fiber. In this case, the core sample would overestimate the value of the hay. Therefore, the composition of the wasted feed should be estimated.

Table 4.4.   Effect of Stage Maturity on Protein
Content and Dry Matter Digestibility of Alfalfa

Date of Harvest	Stage of Maturity	Crude Protein (%)	Dry Matter Digestibility (%)
May 31	Vegetative	21	66
June 5	Early bud	20	65
June 10	Bud	19	63
June 15	Early bloom	18	61
June 20	Quarter bloom	17	60
June 25	Half bloom	16	58
June 30	Full bloom	15	57

[a] Adapted from data of M. J. Anderson, *Utah Sci.* 1975.
[b] Dry matter basis.

In the future the use of infrared methodology may greatly reduce the labor, cost, and time required for analysis. A specific factor in a given product will reflect certain wavelengths of infrared light in a specific way for the determination of protein, fiber, fat, calcium, phosphorus, and many other nutrients. Some scientists feel that more testing is needed before infrared analysis can be widely recommended, but the method has tremendous potential.

If analyses are not available, or if only small amounts of hay are being purchased, the hay can be evaluated by using book values in the National Research Council's *Atlas of Nutritional Data* and by visual appraisal.

The amount of leafiness is a good indication of nutritional value because leaves have more nutritional value than stems. Early cut alfalfa is about half leaves and half stems. If the plant is too mature when cut or damaged by weather and excessive handling during harvest, leaf loss can be 70 percent or greater. If the leaves are not firmly attached to the stems, they can be lost during feeding. Shattered leaves are not readily consumed by horses. Hay that is too dry when baled often has excessive shattering of leaves.

Table 4.5. Suggested Stage of Growth for Harvesting Hay Crops

Crop	Stage to Harvest
PERENNIAL GRASSES:	
Bluestem, introduced and native	Boot to early bloom
Coastal Bermudagrass	14–16 in. height (maximum 4 weeks growth)
Common Bermudagrass	Early bloom
Fescue	Boot to early head
Johnsongrass	Boot
Lovegrass	Boot
Orchardgrass	Boot to early head
Smoot bromegrass	Early to medium head
Timothy	Boot to early head
ANNUAL GRASSES:	
Millet, Pearl or cattail	Boot
Oats	Late milk to early dough
Ryegrass	Early bloom
Sorghum, forage	Bloom to soft dough
Sudan varieties and sudan hybrids (including sorghum almum, and sorgrass)	Boot
LEGUMES:	
Alfalfa	Full bud
Birdsfoot trefoil	$1/4$ bloom
Clover (crimson, hop, red, white, Persian, and arrowleaf)	$1/4$ to $1/2$ bloom
Peanut	Harvest to retain maximum leaves
Vetch	Early bloom

The ratio of stem to leaf is important. The diameter of the stem varies with the type of hay. Birdsfoot trefoil, for example, will have a much finer stem than sweetclover. As a general rule, fine stems are more nutritious than course stems.

Hay should have a pleasant odor. Musty or offensive smells suggest mold, and hay should be mold-free. Horses are believed to be more susceptible than ruminants to moldy feed; they can suffer from botulism, vitamin K deficiency (moldy sweet-clover disease), liver damage, aggravation of heaves, and other problems. Hay should be taken apart and examined closely for mold or dust, especially if it has been exposed to dampness or heat. Mold might also harm horsemen. A person working with moldy hay can have as many as 10 million thermophilic bacterial spores deposited in the lungs per minute. Dr. L. Easson (1980) wrote that after initial exposure to the spores some people develop "farmer's lung" because of an allergic reaction. Several hours after the exposure there could be a flu-like condition with sweating, coughing, aches and pains, and general weakness. Repeated exposure can cause progressive scarring of the lung tissues, leading to a gradual loss of lung efficiency and permanent damage. Dr. Easson recommended that moldy bales never be opened in a confined space. He said that simple gauze or paper masks offer little protection, but a properly fitted full face respirator with a canister-type filter can be effective.

Hay should be free of weeds, grain stubble, and any other foreign material that would reduce nutritive value or palatability.

The inside of the bale of properly cured hay should be green. Brown or discolored hay usually means weather damage, excessive heating after baling, or that the plant was too mature at harvest. Of course, there will be some discoloration of the outside of the bales that are exposed on the sides of the stack or mow.

Fertilization rate can also influence hay quality, particularly in grass hay. One study at Texas A and M University demonstrated that the crude protein content of unfertilized tall fescue and smooth bromegrass could be increased from 11 percent to 17 percent when 150 pounds of nitrogen per acre was added. Estimated dry matter digestibility increased from 57 percent to 60 percent.

In studies at Cornell, chopped hay did not infleunce digestibility nor increase intake, but it took extra labor and equipment to prepare. In addition, horses fed chopped hay mixed with grain consumed the grain at about one third the rate they took to consume long hay and grain separately. Remember this if your horse is greedy! A mixture of chopped hay and grain could also be used with group-fed young animals to prevent overeating. The preference of the horse for long or chopped hay may be influenced by dietary history.

LEGUMES

Alfalfa is the number one hay in the United States; it is often called the "Queen of Forages." The USDA reports that 40 percent of all hay produced in the U.S. is alfalfa or alfalfa-mixed. The biggest producers are Wisconsin, Minnesota, California, Iowa, and Kansas. Mature animals at maintenance do not need the high levels of protein and calcium in alfalfa, and when fed it, they excrete high levels of nitrogen and calcium. Mature alfalfa-fed animals may urinate more and the smell of ammonia may be strong in a poorly-ventilated barn because of the additional nitrogen passed. The increased nitrogen excretion does not cause kidney damage.

The high protein, calcium, and vitamin content of alfalfa is particularly useful in the feeding of young animals and breeding animals. Dr. T. W. Swerczek at the University of Kentucky suggests that lactating mares fed excessively nutritious diets such as those containing alfalfa are likely to produce milk containing a high level of fat and corticosteroids. He says that three- to eight-week-old foals drinking such milk are more likely to develop the shaker foal syndrome: sudden onset of severe muscular weakness and tremors, so that the foal often has to be helped to his feet. The condition worsens and the foal is soon unable to stand even with assistance. Death may occur within 72 hours because of respiratory failure. Dr. Swerczek recommends that lactating mares not be fed alfalfa or high-energy feeds such as corn but rather timothy hay and oats. However, I feel that more research is needed before this can be considered conclusive. Many horse owners have fed alfalfa to mares for years without difficulty.

Some horse owners do not like alfalfa hay for other reasons. It is true that alfalfa is a controversial feed; to help you make up your own mind, I am listing some additional advice, with studies where relevant:

1. Improperly harvested alfalfa hay may be quite dusty and irritating to horses, particularly those with respiratory problems such as heaves.

2. It is much easier to overfeed horses alfalfa than timothy

because of alfalfa's high digestible energy content. In fact, some horses can become quite fat on a high-quality alfalfa hay diet even if they receive no grain.

3. Some owners feel alfalfa is too hot. A fat horse is more likely to sweat than a thin horse, and perhaps this has led to the feeling by some horsemen that horses fed alfalfa are likely to sweat more profusely than those fed timothy. However, studies with Cornell polo ponies on similar work schedules showed no difference in the amount of sweat produced when the ponies were fed equal amounts of digestible energy from corn-alfalfa rations or oats-timothy rations.

Four polo ponies were fed corn and alfalfa pellets and three were fed oats and timothy hay for four weeks. The horses fed corn and alfalfa were also fed three pounds of timothy hay daily to provide fiber. The diets were then switched and fed to the horses for another four weeks. Thus, each horse served as its own control.

Feed intake was regulated to maintain the horse's weight throughout the eight-week period. Body temperature, pulse, respiration rate, and a blood sample were taken from each animal weekly before and after a varsity polo practice. Practice consisted of a warmup period and two non-consecutive seven-minute chukkers. The blood was analyzed for total plasma protein (TPP) and osmolality, to determine the concentration of electrolytes in the plasma. (Sweating increases concentrations of protein and electrolytes in the plasma.) The horses were also appraised visually for sweat at the end of each practice session, on a scale of one to ten.

The horses had been used for polo for six months prior to the start of the study and were in good physical condition. Their average weight was 1,150 pounds.

No significant differences due to diet were found in any of the criteria used (Table 4.6). The TPP concentration and osmolality increased at the same rate in both groups and the visual sweat scores were similar for both groups. The corn-alfalfa ration did not cause the horses to sweat more. However, it took more pounds of the oats and timothy hay ration to maintain body weight.

We have also fed rations containing 40 percent alfalfa and 60 percent corn to horses doing extended work. One trial was con-

Table 4.6.   Pre- and Post-Exercise Values for Polo Ponies Fed
Corn-Alfalfa or Oats-Timothy Rations

		Corn-Alfalfa	Oats-Timothy
	Alfalfa		
Body temperature (°F)	Pre	100.6	100.3
	Post	103.2	103.2
Pulse (beats/minute)	Pre	41.0	41.9
	Post	83.1	86.7
Respiration (breaths/minute)	Pre	15.7	16.2
	Post	82.0	88.2
Total plasma protein (mg/100 ml)	Pre	7.10	7.16
	Post	7.25	7.36
Osmolality (milliosmols)	Pre	282	277
	Post	296	292
Sweat score		2.94	3.64

ducted with 6 Thoroughbreds and two trials were conducted with 16 Arabians. The Thoroughbreds were ridden on 4 rides of 37 miles at a rate of 6 miles per hour and the Arabians were ridden on 4 rides of 50 miles at a rate of 9 miles per hour. None of the horses appeared to sweat excessively during training or during the rides, and no difficulties could be attributed to the corn-alfalfa ration.

4. Alfalfa has a wide Ca:P ratio. The average is about 6:1, but some hays—particularly those raised in areas with low phosphorus soils, such as the western States—may have ratios as high as 15:1. Mature horses appear to be able to handle ratios as wide as 6:1 provided the level of phosphorus is adequate, but ratios of not more than 3:1 are recommended for young horses. The Ca:P ratio of the diet can be changed by the addition of grain and mineral supplements.

5. The availability or absorbability of calcium in alfalfa by the horse is also under question. Studies by Dr. George Ward and co-workers (1979) demonstrated that some alfalfa has a high content of oxalate, an organic acid. The oxalate combines with calcium and forms crystals. The researchers suggested that 20 to 33 percent of

the calcium in alfalfa can be in the form of calcium oxalate and is not readily absorbed by cattle. Further experiments are needed with horses, but earlier studies indicate that horses could absorb about 70 percent of the calcium from alfalfa.

6. Another potential problem with alfalfa hay is contamination with blister beetles (*Epicauta species*). Dr. T. R. Schoeb and Dr. R. J. Panciera (1979) from Oklahoma State University have reported 21 cases of beetle poisoning. They stated that the most consistent clinical signs were colic, fever (up to 105°F), and increased pulse and breathing rate. Sweating and soft feces were noticed occasionally. Several horses were seen repeatedly splashing their muzzles in water without drinking. Hypocalcemia (low blood level of calcium) was observed in two thirds of the horses. Only 6 of 21 survived. Of the 15 that died, 11 died within 2 days after the first signs of illness. Necropsy findings included changes in the gastrointestinal and urogenital tracts and in the heart. The epithelium (first layer of tissue) was missing in parts of the stomach of several horses. The intestines were inflamed and tissue contained more water than usual. The lining of the urinary bladder was ulcerated. The heart muscle suffered severe damage.

Drs. Schoeb and Panciera suggested that the repeated immersion of the muzzle in water could be a response to irritation to the lining of the mouth.

The Oklahoma workers did not find beetles in any feedstuff other than baled alfalfa hay. The beetles feed in swarms in alfalfa. Schoeb and Panciera concluded that modern harvesting of alfalfa by cutting and crimping in a single operation sometimes results in large numbers of beetles in individual bales or even a single flake.

Most of the cases of beetle poisoning have been from the South or Southwest. I am not aware of any cases from the Northeast. One species of the beetle (*Epicauta occidentalis*) has been found in Nebraska, Kansas, Colorado, Oklahoma, Texas, and Louisiana. Another species (*E. lemniscata*) can be found in the above-mentioned states and east to New Jersey and Florida. A third species (*E. vittata*) is found from Iowa east to Maine and New Jersey. The black blister beetle (*Epicauta pennsylvanica*) is common in the Northeast but usually doesn't appear until August, and seldom in alfalfa fields. It seems to prefer goldenrod.

There doesn't appear to be a specific or satisfactory treatment for beetle poisoning. The toxic material is probably cantharidin, a powerful irritant and blistering agent. Cantharides have also been reported to cause death in humans ingesting insects or extracts of insects such as the blister beetle, blister fly, or Spanish fly. But a specific antidote has not been developed. Doses of mineral oil might help protect the lining of the intestines and decrease absorption of the toxic compound.

The best course seems to be prevention by examining the hay before feeding, particularly in areas with a known large population of beetles. The beetle is about $^1/_4$ to $^3/_4$ inch long and may be entirely black, black with orange stripes, gray, or yellowish tan with or without black spots.

Unfortunately, there isn't a practical way of treating the alfalfa to keep the beetle away, although there are several insecticides that will kill it.

In some areas of the United States, blister beetles are even considered to be a helpful insect, because during the larval stage they feed on the egg pods of various species of grasshoppers. Other species, however, are honey thieves. They crawl on unsuspecting bees and are carried to the hive, where they eat the eggs and stored food.

Birdsfoot trefoil can be an excellent hay for horses. The nutritive value of properly harvested material is similar to that of alfalfa. It may be difficult to establish a stand, but once established, trefoil can be high-yielding and may grow on poorly drained areas not tolerated by other legumes. It usually has a finer stem than clover hay. Some reports suggest that horses are not particularly fond of trefoil hay, but I have talked to horse owners who say that their horses prefer trefoil if given a choice.

Several varieties of trefoil are available. Some have been bred primarily for hay, others for pasture.

Red clover hay, if properly harvested, has a nutritional value similar to that of alfalfa. Many horsemen in the 1800s and early 1900s were prejudiced against red clover hay because they thought it caused heaves. Average clover hay is usually more dusty than alfalfa hay. And there is no doubt that dusty red clover hay aggravates heaves in a horse; however, it does not cause heaves. Several studies have demonstrated that horses relish good quality red clover hay.

Crimson clover is tolerant of humid weather but not of severe cold; therefore it is much more common in the Southeast than the North. It appears to be higher in fiber than red clover and must be cut early to remain palatable. Dr. W. A. Wheeler (1950) wrote, "As crimson clover reaches maturity the hairs of the heads and stems become hard and tough. When it is grazed continuously or when it is fed as hay at the stage, large masses of the hair are liable to form into hair balls in the stomach of horses and mules, occasionally with fatal results."

Sweet clover is not widely used for hay. It is perhaps more noticeable as a roadside weed than as a hay crop, and bees love it. Although sweet clover is very resistant to cold and drought, it matures early, has a higher content of stems to leaves than alfalfa, and the stems are particularly slow in drying. It is difficult to obtain well-cured hay without significant loss of leaves. Proper curing is essential because otherwise mold can develop; moldy sweet clover has long been known as toxic, mainly with cattle.

Dicumarol is produced by the action of certain molds on coumarol, a natural constituent of sweet clover. Dicumarol competes with vitamin K in the clotting process—hence, moldy sweet clover causes failure of the blood to clot. There could be hemorrhages into tissues and severe blood loss after injury, surgery, or parturition.

Although sweet clover is not a major crop in the United States, it is grown extensively in some states with cold climates such as North Dakota and in Canada. Dr. B. P. Goplen (1980) recently reported that sweet clover produces more forage per acre than any other legume grown in western Canada, including alfalfa. It is valuable for soil improvement, silage production, hay and pasture, and can be grown in saline soils found unsuitable for cereal crops and most other legumes. There is a white-flowered cultivar that contains low levels of coumarin; a yellow-flowered cultivar with low levels of coumarin will soon be available.

A report from the University of Saskatchewan by Dr. G. K. McDonald (1980) described the case of one horse suffering from sweet clover toxicity. A 6-year-old Percheron mare had a nosebleed for 24 hours, and the blood took a long time to clot. The mare had been fed free-choice weathered sweet clover hay. Treatment of a single whole blood transfusion and four daily intravenous injections of vitamin K_3 was successful. Dr. McDonald pointed out that it is difficult to produce mold-free sweet clover hay because of the succulent nature of the plant and the coarse, heavy stems. There also appears to be an increased incidence of the disease in wet haying seasons.

Not all moldy sweet clover contains dicumarol; varieties of the clover vary in their cotent of coumarol and toxic potential. Interestingly, studies on the toxicity of moldy sweet clover led to the development of rodent poisons containing dicumarol.

Lespedeza was much more popular in the past than it is now. It was thought to be a legume that could be used to produce a good yield of high-protein hay on poor soils and even prove as good as alfalfa hay. But Dr. E. I. Robertson (1981) reported that feed manufacturers now have only a slight interest in lespedeza. Dr. M. Bradley (1981) pointed out that lespedeza has not yielded as expected, is not competitive with other legumes, and has been troubled by disease. When overmature and overdried, it has tough wiry stems and is subject to severe leaf shatter.

GRASS HAY

Oat hay can be excellent for horses and is a major annual forage in western United States and Canada. About seven percent of the oat acreage planted in North Dakota each year is harvested as forage. Dr. Erickson and co-workers (1976) at the University of North Dakota reported that medium- to late-maturing oat varieties such as dal, lodi, kota, and cayuse yield more forage than earlier-maturing varieties such as nodaway 70, otter, and chief. If the plant is harvested too early, yield will suffer. On the other hand, as the plant matures, digestible energy, protein, phosphorus, and potassium concentration decrease. Oats should be cut for forage when the grains are at the late milk to early dough stage for maximum forage nutrient production (Table 4.7). The North Dakota workers

Table 4.7. Effect of Growth Stage at Harvest on Nutrient Concentration in Oat Hay

| | PERCENTAGE OF DRY MATTER | | | |
Stage	Protein	Phosphorus	Potassium	Calcium
50% headed	15.6	0.23	2.3	0.32
Fully headed	14.9	0.23	2.2	0.33
Milk	12.2	0.19	1.6	0.27
Soft dough	12.2	0.18	1.6	0.28
Hard dough	11.5	0.18	1.5	0.31
Mature	10.8	0.17	1.2	0.28

Erickson et al., 1976, North Dakota Farm Research.

also suggested that oat hay can contain toxic levels of nitrates and nitrites and should thus be analysed before being fed. Nitrate can be converted by gastrointestinal bacteria into nitrite. Nitrite causes the conversion of hemoglobin into methemaglobin, which is incapable of carrying oxygen. Oxygen deficiency is a killer. Cattle are much more susceptible to nitrate problems than horses, but scientists at the University of Wyoming have demonstrated that the feeding of high levels of nitrate can be fatal to horses. The toxic level of nitrate for horses has not been determined. Concentrations of up to 0.66% nitrate ion (dry matter basis) are safe for nonpregnant cattle, but pregnant cattle should be limited to 0.44% nitrate— and these levels should be well tolerated by horses.

Timothy is widely adapted, tolerates cold weather, produces well in early spring, is easy to establish, and grows under a wide range of soil and climatic conditions. Timothy is the hay most preferred by horse owners in many areas of the United States. When properly harvested it is an excellent feed. Although timothy does not contain as high a level of protein or calcium as legumes, early cut timothy can contain 12 percent or more protein and at least 0.5 percent calcium. Unfortunately, most timothy hay fed to horses contains much lower levels of these nutrients, perhaps 8 percent protein and 0.3 percent calcium.

Good quality timothy has little dust. Although horses will usu-

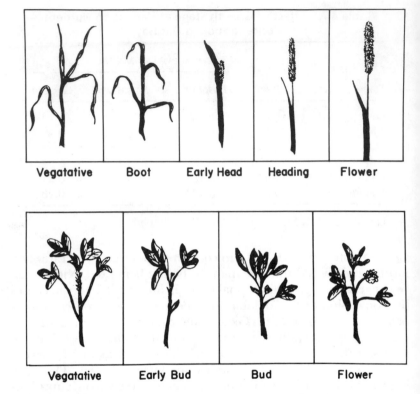

Figure D.   Stages of growth of timothy (upper) and alfalfa (lower). (Courtesy R. Seaney, Cornell University)

ally select alfalfa if given a choice, they like good quality timothy and eat it readily. (See Figure D for growth stages of timothy and alfalfa.)

Many horsemen say that the feces of timothy-fed horses are firmer than the feces of horses fed legumes such as alfalfa. But studies by Dr. Paul Fonnesbeck show that once horses adjust to an alfalfa diet, the feces become firm.

Coastal bermudagrass is a warm-season perennial grass. It is the most important grass of the South and Southeast, even though it was once considered a serious pest by cotton and tobacco farmers.

Now it is used for hay, pasture, home lawns, and golf courses. Bermudagrass is heat- and drought-tolerant and useful for erosion control. Dr. W. O. Wheeler (1950) suggested that burmudagrass is good for hay production in the humid Southeast because no other hay plant can be cured so quickly and handled so easily.

Smooth bromegrass is a common grass in pasture and hay fields of the North, particularly in the Corn Belt states and the Northeast. It is frequently seeded with alfalfa. Smooth bromegrass is slow to establish but is a long-lived and productive grass.

Silage

Silage is the material produced by controlled fermentation of high-moisture herbage. The process is self-limiting; when stored under anaerobic (no oxygen) conditions, the production of lactic acid eventually stops the fermentation process. Any high-moisture forage can be ensiled. The whole corn plant, grasses, and legumes such as alfalfa are the crops most commonly ensiled in the United States. Silage has several advantages. The ensiled corn plant yields a great amount of digestible energy per acre that would be impossible to harvest by any other practical method. Weather losses of grasses and legumes can be avoided, and the feeding of silage— unlike hay can be mechanized readily. Silage is much easier to use than hay in complete rations, but it must be consumed soon after removal from storage or else spoilage will occur.

Silage is an excellent feed for ruminants; the amount used in dairy cattle rations has steadily increased.

F. B. Morrison (*Feeds and Feeding*) reports that many farmers feed silage to draft horses and mules. It must be of good quality, free of mold, and should not replace more than one third to one half the hay feed. He further suggests that silage not be fed to horses at hard work but to idle horses, especially idle brood mares.

It was recently reported that the mares at the Irish National Stud have been self-fed grass silage for several years without any difficulty. But horses are apparently much more susceptible than ruminants to toxins in moldy silage. Three horses fed silage were recently brought to the Cornell Large Animal Clinic. Two died and the third required a long recovery period. Unfortunately, the potential toxicity of a silage sample cannot be ascertained reliably by just looking at it. The risk of toxicosis must therefore be balanced against the amount of feed dollars saved.

Although silage is aromatic, the smell apparently does not offend horses. Usually they will eat it readily. (Sauerkraut is an example of aromatic silage consumed by humans.)

Corn silage is lower in protein and calcium content than grass or legume silage but is likely to be higher in digestible energy because of grain content. Corn silage may contain 30 to 40 percent dry matter, with 7 to 10 percent crude protein, 24 to 34 percent acid detergent fiber (ADF), 0.25 percent calcium, and 0.21 percent phosphorus. Grass silage may contain 25 to 40 percent dry matter, with a crude protein content of 7 to 13 percent protein, 35 to 45 percent ADF, 0.6 percent calcium and 0.2 percent phosphorus. Legume silage might contain 15 to 22 percent protein, 32 to 45 percent ADF, 1.35 percent calcium, and 0.25 percent phosphorus.

Straws

Straw, or the residue after the grain or seed has been harvested, is generally much higher in fiber and lower in digestible energy and protein than hay. It can, however, be incorporated into a ration as a source of fiber and can also be treated to improve digestibility. Efforts are being made to develop methods of processing straw to improve utilization by ruminants and horses.

Straws—in addition to being low in protein and digestible energy—contain little or no vitamin A activity or B vitamins. They have very low levels of phosphorus and about two thirds the calcium content of grass hay. So straw may be considered primarily a "filler," although mature horses that are not working have been maintained by diets high in straw. In Cornell studies, ponies fed diets containing 65 percent oat straw, 5 percent soybean meal, 20 percent ground corn, and 10 percent molasses maintained their weights during a 4-month feeding period.

Treating straw with ammonia can increase the digestibility of dry matter to about that of hay. Other studies, such as those by Dr. W. A. Schrug (1981), have shown that treatment with NaOH is also beneficial. The chemical treatments disrupt the cell walls, perhaps by dissolving lignin and by "swelling" cellulose, thus making the material more digestible.

Workers at Oregon State University have maintained ponies on diets containing 60 percent ryegrass straw. Treating the ryegrass straw with acid and inoculating with yeast improved the digestibility of the dry matter.

Presently the chemical treatment of straw is not economical and the residue of the NaOH treatment causes disposal problems. But it is highly likely that treated straws will have a much more important place in livestock feeding in the future.

Pasture

Good pasture provides open space for horses to run and exercise freely in sunlight and fresh air. It reduces labor and can be one of the most nutritious as well as inexpensive sources of feed for horses. Good quality pasture, containing a mixture of legumes and grasses, plus trace-mineral salt (free choice) and water can supply the nutrients needed by most classes of horse. Exceptions would be hard-working horses and young horses in which a fast rate of gain is desired. It is much easier to be a good horse nutritionist when good quality pasture is available. "Dr. Green" (good pasture) can cure a lot of problems. And yet pasture management is neglected on many horse farms.

County agents and extension specialists can provide a wealth of information about pasture management. They are familiar with the conditions in your locality. In addition, there are numerous extension bulletins available on pasture management, such as:

1. *Establishing forage crops.* Dept. of Agronomy, University of Kentucky, 1981.
2. *Nitrogen fertilization of orchardgrass pasture.* Bulletin, College of Agriculture, Washington State University, Dec. 1979.
3. *Evaluation of alfalfa varieties for hay production in northwest Texas,* Texas Agricultural Exp. Sta., Oct. 1979.

4. *Kentucky forage variety trials.* Report Kentucky Agr. Exp. Stat., April 1979.
5. *Irrigated pastures for horses in eastern Washington.* Bulletin, Wash. State Univ. Coop. Ext. Serv., 1979.
6. *Non-irrigated pastures for horses in eastern Washington.* Bulletin, Wash. State Univ. Coop. Ext. Serv., 1979.
7. *Horse pasture.* Dept. of Agronomy, University of Kentucky, 1981.
8. *Pasture improvement and management for horses.* Information Bulletin, 171, Cornell University, 1981.
9. *Alfalfa management in North Dakota.* no. R-571. North Dakota Coop. Ext. Service., Jan. 1981.
10. *Kansas rangelands—their management.* Bulletin, 662, Kansas State Univ., Oct. 1978.
11. *White clover for South Carolina.* Bulletin, 592, Clemson Univ., Aug. 1976.

To ensure sound pasture management, you should:

1. Test soil to determine how much lime and fertilizer are needed.
2. Clip to reduce weeds and to stimulate new growth of the pasture plants.
3. Prevent overgrazing. Most pasture species should not be grazed closer than about two inches.
4. Avoid undergrazing. This can cause loss of legume stands and lowered quality because the grass plants become too mature.
5. Scatter manure piles.
6. Remove weeds—particularly toxic plants.
7. Keep horses off pasture if the ground is too wet. Hooves can really plow up a pasture and destroy plants, and the holes can be dangerous when the ground dries.
8. Maintain a mixture of legumes and grasses to improve quality yield. This results in less heaving of soil and, therefore, less exposure of roots. (Exposure of roots increases susceptibility to disease.)
9. Select the proper plant species to improve yield of nutrients.

Scattering of fecal piles with a flexible tine harrow. (Courtesy Fuerst Brothers, Rhinebeck, New York)

Table 4.8 gives a summary of advantages, disadvantages, and concerns about various species of pasture plants. The mixture of plants should be selected according to their tolerance of the growing conditions in your area, so do not hesitate to consult your local county agent or extension specialist.

Cattle and sheep can be used to help control spot grazing and parasites, reduce amount of clipping required, and provide more economical pasture for horses. If used judiciously, the cattle or sheep can be added without decreasing the number of horses that the pasture can carry. Thus the cattle not only improve the pasture but provide a saleable product. They can be rotated on the pasture with horses, although some farms run them together.

Table 4.8 contains information about several of the more common forage plants. Many more plants, however, are being studied at experiment stations throughout the United States in an effort to improve pasture productivity. Frost blue lupine, limpograss, pangola digitgrass, hairy indigo, buffelgrass, prairie grass, bluestem, blue panicgrass, Kleingrass, Rhodesgrass, and arrowleaf clover are just a few examples.

Sudan grass is not the only plant that can cause prussic acid poisoning. Many plant materials such as arrow grass, Johnson grass, wild black cherry, chokecherry, and pits of peaches, apricots, and plums contain cyanogenetic glycosides. These, when hydroloyzed by an enzyme during ingestion, yield prussic acid (hydrocyanide) and can cause cyanide poisoning.

Laetrile—claimed by some to be a cure for cancer—is made from apricot pits and has been reported to contain a cyanide-releasing substance.

Bitter almonds may also contain some hydrocyanide and large doses of this may cause death within a few minutes. But animals may linger for an hour or more when smaller doses are ingested. Signs of poisoning include convulsions, frothing at the mouth, unconsciousness, and gasping. The heartbeat is rapid and weak. Pupils are usually dilated. Involuntary defecation and urination may occur.

In acute cases of poisoning, the venous blood becomes bright red because cyanide acts by preventing the intracellular oxidative process.

Treatment often consists of sodium nitrite and sodium thio-

Table 4.8. Brief Summary of Characteristics of Some Pasture Plants for Horses

Species	Advantage	Disadvantage or Concern
	LEGUMES	
Alfalfa	Highly nutritious. High yield. High palatability.	Requires well-drained soils with pH of 6.5 to 7. May not survive long as a pasture because it cannot tolerate heavy grazing or traffic.
Alsike	Can tolerate land that is too wet or acid for red clover.	An early report claimed that alsike could cause liver damage in horses and that horses would not eat alsike unless no other pasture was available.
Birdsfoot Trefoil	Can be productive on poorly drained soil. Persists by natural reseeding.	May be difficult to establish; seedlings lack vigor. Many horses prefer other legumes to birdsfoot.
Ladino Clover	More productive than common white clover.	Not drought-tolerant.
Lespedeza	Can tolerate heat better than some legumes.	Not highly productive. Tough stems.
Red Clover	Highly nutritious. Tolerates acid soils better than alfalfa.	May survive only 2 to 3 years as pasture. Cannot tolerate heavy grazing. Mold infection may cause horses to slobber.
White Clover (sometimes called Common or Dutch)	Low-growing, tolerant of overgrazing. Winter-hardy.	Not as highly productive as many legumes. Mold infection may cause horses to slobber.
	GRASSES	
Bahiagrass	Warm-weather grass. Adapted to sandy, infertile soils. Retains quality later in season than Bermudagrass.	Yield not as great as Bermudagrass.

Table 4.8. (Continued)

Species	Advantage	Disadvantage or Concern
Coastal Bermudagrass	Warm-weather grass. Well adapted to most soils. Dense turf, can tolerate heavy traffic. Most important summer pasture in southern U.S.	Not adapted to poorly drained soils.
Dallisgrass	Warm-weather grass.	May have ergot problem. Not suited to sandy soils.
Kentucky Blue-grass	Considered the No. 1 pasture grass for horses in many states. Palata-ble, tolerates close graz-ing, winter-hardy. Forms smooth, dense sod; controls erosion.	Somewhat slow to establish. Some varieties dormant in hot weather, cannot be grown successfully in Deep South.
Orchard Grass	Produces early spring growth. More drought- and heat-tolerant than bluegrass or timothy.	Palatability decreases rapidly as plant matures. Not as win-ter-hardy as timothy. Does not form dense sod.
Reed Canary Grass	Does well on wet land. Will grow on land sub-ject to overflow for pe-riods that would kill other grasses. Tolerates drought well.	As plant matures, alkaloid content increases, greatly re-ducing palatability. Seeds lose germibility readily during stor-age. Do not use old seed.
Smooth Brome-grass	Important pasture in the North. Forms dense sod, makes good re-growth. Grows on a wide variety of soil types. Retains palatabil-ity at later stages of sea-son than most grasses.	Slow to establish. Grows best in well-drained soil. Not toler-ant of overgrazing.
Sudan Grass or Sudan Grass Hy-brids	Not recommended for horse pastures because of reports that it can cause cystitis (infection of the urinary bladder) and abortion. Affected mares may have vaginal and ure-	

Table 4.8. (Continued)

Species	Advantage	Disadvantage or Concern
	thral irritation. Yellowish, sticky, granular fluid accumulates in bladder. Death may be caused by kidney infection. (Sudan grass hay does not seem to cause the problem.) There is also danger of prussic acid poisoning. New growth after trampling or frost contains a high level of prussic-acid-yielding compounds.	
Tall Fescue	Long-lived. Tolerates heavy traffic and hot weather. Produces high yield of dry matter per acre.	Low palatability. If consumed by pregnant mares during eleventh month of gestation, may cause prolonged gestation, cause thickened placenta, abortion, weak foals, or lack of milk production.
Timothy	Easy to seed and establish. Produces well in early spring. Grows under a wide range of soil and climatic conditions.	Bunch grass—a sod former and susceptible to weeds.

sulfate, which detoxify the blood by stimulating the formation of cyanomethemoglobin and thiocyanate.

Prussian blue, a complex of cyanogen compounds, is used in paints, dyes, and inks. It was discovered in 1904 by a resident of Berlin named Diesbach. Later, another chemist isolated hydrocyanide from Prussian blue and named it prussic acid. Hydrocyanide is colorless.

Often, pasture may need renovation. The use of herbicides and no-till drills makes this more economical, especially as the drills can be used to plant forages on sites that cannot be plowed because of slope, erosion hazard, or stones.

A mixture of legumes and grasses is usually best, but careful

John Deere 251 Powr-Till® Seeder is a large capacity, pull-type no-till drill designed for rangeland renovation. (Courtesy John Deere)

management is required to maintain the mixture. In warmer climates, warm-season grasses are more tolerant than legumes to pathogens, heat, and drought. The grasses can withstand closer grazing than the legumes, particularly because animals will often prefer legumes to grasses. Proper fertilization, liming, and clipping can help maintain the desired mixture.

The amount of pasture land required per horse depends on many factors. High-quality, well-managed pasture might easily support one horse per acre. Pasture in which fertilizer, lime, or water is limited may require 10 or more acres per horse. A survey conducted in Kentucky indicated that an average of 4 acres was used per horse in that state; this would also include some hay land.

Although the horse is primarily a grazer, he may browse on many different plants. For example, Dr. R. Keiper, who has spent several years studying feral ponies, reported the ponies on islands in Chincoteague Bay at least occasionally eat poison ivy. Fortunately, poison ivy is not toxic to equids. Perhaps even more surprising is that Chrismas holly *(Ilex aquifolium)* was raised as a fodder for cattle, sheep, and horses in ancient England. Of course, not all plants eaten by horses are nutritious; many are even toxic. There have been numerous cases of poisoning in horses caused by the ingestion of Japanese yew, for example.

All horsemen should be aware of poisonous plants. If in doubt, simply consult the bulletins of your local experiment stations. There are also several books available such as *Poisonous Plants* by J. Kingsbury.

Grains

Grains are the most commonly used high-energy sources for horses. Grains contain about 50 to 60 percent more digestible energy per pound than hay because they are much more efficiently digested. Almost any of the grains can be fed to horses if the characteristics of the grain are considered and adjustments are made in the feeding program.

This horse is getting good care, but a few mouthfuls of the Japanese yew (bush in the foreground) would be toxic. (Courtesy Equine Research Program, Cornell University)

CORN

Corn is the king of grains fed to livestock in the United States. Production has almost doubled in the last 15 years and now is at least 10 times that of any other grain except wheat. It is almost four times that of wheat (and, of course, most of the wheat is raised for human consumption).

Many horsemen do not like to feed their horses corn because they feel it is a "hot food." True, corn is a concentrated source of energy and must be fed carefully. One quart of corn provides as much digestible energy as two quarts of oats, so it is easier to overfeed corn than oats. But when both grains are fed to provide equal amounts of net energy, oats actually provide more "heat" because of a higher fiber level.

Oats have two primary advantages over corn: a higher fiber content, making it a "safer" feed, and a higher protein content. Corn has three primary advantages over oats: a lower cost per unit of digestible energy, more consistent quality, and a higher content of vitamin A activity. In fact, corn is the only common grain that contains any vitamin A activity.

Although the quality of corn is not as variable as that of oats, poor-quality corn should be avoided. In a study at the University of Georgia, pigs were fed U.S. No. 2 corn (good quality—see Table 4.9 for description of grades) or U.S. No. 5 (poor quality—contains kernels that have failed to reach mature size probably because of hot, dry weather during grain filling stage, some dark, broken kernels, weed seed, and even bits of corn stalks). The pigs fed the U.S. No. 5 corn required 15 percent more feed and gained weight at a slower rate than those fed good corn. Similar results would be expected with horses. Poor-quality moldy corn may also contain aflatoxin and toxins produced by *Aspergilli*, causing degeneration of the liver and parts of the brain. Moldy corn may also contain *Fusarium verticillioides*, which can cause leukoencephalomalacia (destruction of the white matter in the cerebral hemispheres of the brain). Other molds may cause feed refusal. Of course, any moldy grain can cause these problems.

Corn can be fed whole, cracked, or on the ear to horses. It

should not be finely ground as it becomes quite dusty. Ear corn is good for horses; it usually forces them to eat the grain more slowly and chew the grain more thoroughly. Some horses may even eat the cobs. Earlier farmers frequently ground the ear to make corn and cob meal for their working horses. The corn and cob meal had a digestible energy content only slightly higher than oats. However, the advent of the picker-sheller has greatly decreased the supply of ear corn.

OATS

Oats have been the favorite grain of horsemen for many years. It has a higher fiber content than corn, barley, or rye, and is thus a much safer feed. Oats also have a higher protein content and quality than corn. Horses fed oats are much less likely to be overfed and are less susceptible to obesity, founder, and digestive upsets.

Unfortunately, cost factors will probably decrease the amount of oats fed to horses in the future. The production of oats has decreased 50 percent and the price has tripled in the last 15 years. Many agronomists predict that the amount of oats produced will continue to decline because the net return per acre is much less than with other grains. As a Mcal of digestible energy from oats can cost 30 percent more than a Mcal of digestible energy from corn, many horsemen will seek cheaper sources of energy. Manufacturers of sweet feeds will be forced to decrease the concentration of oats in their formulas, especially in low-cost varieties.

Another disadvantage of oats is the large variation in quality. Oat kernels can be plump and full of starch, or thin and containing mostly fiber. The fullness of the kernel is greatly influenced by climatic conditions and management practices such as level of fertilization. The weight of oats per bushel indicates quality. More fiber means less weight. The traditional average weight of a bushel of oats is 32 pounds. Anything higher will have a higher digestible energy content; anything lower will have a lower digestible energy content. One coffee can of heavy oats can contain almost twice as

Table 4.9. U.S.D.A. Standards for Grain

CORN			Maximum Limits of		
	Minimum Test Wt. Per Bushel	Moisture	Broken Corn and Foreign Material	Damaged Kernels Total	Heat-damaged
	(lb)	(%)	(%)	(%)	(%)
U.S. No. 1	56.0	14.0	2.0	3.0	0.2
U.S. No. 2	54.0	15.5	3.0	5.0	0.5
U.S. No. 3	52.0	17.5	4.0	7.0	0.5
U.S. No. 4	49.0	20.0	5.0	10.0	1.0
U.S. No. 5	46.0	23.0	7.0	15.0	3.0

U.S. Sample: U.S. Sample grade shall be corn which does not meet the requirements for any of the grades or which contains stones, or which is musty or sour, or heating; or which has any commerically objectionable foreign odor; or which is otherwise of distinctly low quality.

BARLEY	Minimum Limits of		Maximum Limits of				
	Test Wt. Per Bushel	Sound Barley	Total Damaged Kernels	Heat-Damaged Kernels	Foreign Material	Broken Kernels	Thin Barley
	(lb)	(%)	(%)	(%)	(%)	(%)	(%)
U.S. No. 1	47.0	97.0	2.0	0.2	1.0	8.0	10.0
U.S. No. 2	45.0	94.0	4.0	0.3	2.0	10.0	15.0
U.S. No. 3	43.0	90.0	6.0	0.5	3.0	15.0	25.0
U.S. No. 4	40.0	80.0	8.0	1.0	4.0	20.0	35.0
U.S. No. 5	36.0	70.0	10.0	3.0	6.0	30.0	75.0

much digestible energy as a can of light oats. So it is best to feed by weight, not by volume.

Oats are often run through a machine which clips off the pointed end of the hulls and lowers fiber content while increasing

Table 4.9.   U.S.D.A. Standards for Grain (Continued)

MINIMUM LIMITS OF			MAXIMUM LIMITS OF		
OATS	Test Wt. Per Bushel	Sound Oats	Heat-Damaged Kernels	Foreign Material	Wild Oats
	(lb)	(%)	(%)	(%)	(%)
U.S. No. 1	34.0	97.0	0.1	2.0	2.0
U.S. No. 2	32.0	94.0	0.3	3.0	3.0
U.S. No. 3	30.0	90.0	1.0	4.0	5.0
U.S. No. 4	27.0	80.0	3.0	5.0	10.0

		MAXIMUM LIMITS OF		
GRAIN SORGHUM	Minimum Test Wt. Per Bushel	Moisture	Heat-Damaged Kernels	Broken Kernels and Foreign Material
	(lb)	(%)	(%)	(%)
U.S. No. 1	57.0	13.0	0.2	3.0
U.S. No. 2	55.0	14.0	0.5	8.0
U.S. No. 3	53.0	15.0	1.0	12.0
U.S. No. 4	51.0	18.0	3.0	15.0

bushel weight. Bleached oats may appear more attractive, but bleaching may slightly decrease the nutritive value. Crimping or rolling of oats improves the digestible energy content by only 5 to 7 percent for horses with sound mouths. Crimping may be economical for horses under a year of age and for older horses with poor teeth. Whole oats can be stored for much longer periods than crimped oats.

GRAIN SORGHUM

Grain sorghum is raised primarily in the drier areas of Texas and in the Plains. Total production is about 10 percent that of corn. Grain sorghum is well adapted to semiarid climates because it has an extensive root system and modest leaf area. There are many varieties such as milo, kafir, hegari, feterita, and others. Nutritive value is similar to that of corn. Grain sorghum is a high-energy, low-fiber feed, so be as cautious with it as you are with corn. Over-feeding can cause founder and severe—or fatal—digestive problems such as enterotoxemia. But, if properly fed, grain sorghum can be an excellent grain for horses. It should be processed for efficient utilization because of the small size of the grain and the hard coat. Scientists at Texas A & M University have found that the digestibilities of dry matter, crude protein, and starch of micronized heteroyellow grain sorghum were similar to those of crimped grain. They also reported that Quarter Horse foals grew at about the same rate and that feed efficiencies were similar when fed rations containing crimped oats, micronized oats, crimped sorghum, or micronized sorghum.

Some varieties of grain sorghum contain tannic acid, which can greatly reduce palatability. If possible, therefore, test samples on a few horses before buying large amounts.

Interest in grain sorghum production is increasing because water conservation is becoming more critical. Improved varieties of sorghum have increased yields, nutritional value, and insect resistance.

BARLEY

Barley is an excellent grain for horses. It is a good compromise between oats and corn since the digestible energy, fiber, and protein contents fall somewhere in between those of corn and oats. However, barley is much easier to overfeed than oats and is usually not as economical as corn. The amount of barley produced in the United States—mainly in North Dakota, Idaho, Colorado, Minnesota, and the West Coast—is only about 5 percent that of corn.

Barley grown on the West Coast usually contains a lower level of digestible energy and protein and a higher level of fiber than barley grown inland.

Crimping or rolling barley significantly improves its utilization by horses and is recommended.

RYE

Rye is not fed in large amounts to horses. Production has decreased greatly in the last 15 years and is only 0.3% of corn. Moreover, much of that rye is used for human consumption in whiskey and bread; thus rye is seldom an economical source of grain for horses. Horses don't like it very much, so a grain mixture should never contain more than one third rye. It should be processed because the whole, small, hard kernels are difficult to digest. Rye straw is highly prized as a stuffing for horse collars.

Rye may be infected with the fungus *Claviceps purpuria*, which produces a compound called ergot. This fungus growth replaces the grain kernel. It is hard and dark colored, may be two to four times the length of the kernel, and may resemble a rooster's spur in shape (*ergot* is from the French word for "spur.") The fungus can attack any of the small grains and several grasses such as fescue, but rye is one of the most susceptible plants. Ergot stimulates the smooth muscles. It can cause blood supply to the tail, limbs, and ears to be so reduced that dry gangrene develops and the extremities may drop off. Pregnant animals often abort.

Rye bread was the prime staple of the poor in the Middle Ages and the rye was frequently contaminated with ergot. The "Holy Fire of St. Anthony" of the Middle Ages was probably gangrenous ergotism, when the limbs of affected people were "burnt up" or died because of circulation loss. The disease became associated with St. Anthony because his Order took care of the afflicted. The disease was believed to be a sign of divine wrath for breaking the truce with God which restricted fighting to Mondays, Tuesdays, and Wednesdays.

Contaminated rye flour poisoned many humans as late as the early nineteenth century. Even today there are reports (though rare) of abortions in humans caused by ergot poisoning.

WHEAT

Wheat contains about the same digestible energy content as corn, but it is not frequently fed to horses because of the expense. Traditionally horse owners have been advised to feed wheat cautiously, in no more than one third to one half of the grain ration, because it produces a doughy ball in the stomach. Few recent studies have been conducted on the feeding of wheat to horses. Wheat must be processed; steam rolling or steam flaking are better than grinding because grinding causes a more dusty product.

SPELT

Spelt is a close relative of wheat but it has a large fibrous hull and a high fiber content. The digestible energy content is slightly lower than that of oats. Only limited acres of spelt are planted in the United States because other cereals are more productive in most sections.

TRITICALE

Triticale is a cross between wheat and rye, and so is the name: *triticum* (wheat) and *secale* (rye). The cross attempted to combine the quality and uniformity of wheat with the hardiness, vigor, and disease resistance of rye. However, triticale has not proved to be as competitive and little is fed to horses, except for some areas of the world with poor soils.

Other High Energy Sources

MOLASSES

Molasses can be a by-product of sucrose refined from sugar cane or sugar beets, or from the manufacture of dried citrus pulp. It can also be manufactured from the starch of grains. Molasses may contain 70 to 80 percent dry matter; it supplies little or no protein or phosphorus. Molasses from cane sugar may contain significant amounts of calcium and is an excellent source of energy, but in recent years prices have fluctuated greatly. During the last three years the price of cane molasses (per Mcal of digestible energy) has varied from 85 percent to 135 percent of corn. Because molasses improves taste and decreases dust, it is often added to rations to entice animals to eat. "Sweet feeds" were among the first commercial horse rations manufactured. Molasses is also useful in the manufacture of pelleted feeds.

Dried molasses is preferred to wet by many feed manufacturers because less equipment is needed to mix it into the feed.

Some horsemen do not use molasses in the summer because they feel it is a heating feed, but this is not the case when fed at the same level of digestible energy as other feeds. High levels of molasses may cause colic and loose feces, but draft horses and mules have been fed as much as 9 pounds per day without any apparent problems. Of course, feeds with molasses attract flies in the summer.

FATS

Fats are a highly concentrated source of digestible energy but do not provide any protein, carbohydrate, or minerals. Fats are efficiently digested by horses (Table 4.10). They can be of vegetable or animal origin, the latter being more commonly used. Feed-grade animal fat and vegetable fats have also been used in experimental diets for horses doing prolonged work.

Feed manufacturers often add small amounts of fat to rations to decrease dust and to speed up pellet mill capacity. Levels as high as 30 percent have been fed to horses in experiments. Antioxidants should be included to prevent oxidation and avoid the destruction of fat-soluble vitamins such as vitamin A and E. Oxidized or rancid fats are also less palatable.

Table 4.10. Apparent Digestibility of Various
Kinds of Fat by Horses

Kind of Fat	Level in Diet (%)	Digestibility (%)
Corn oil[a]	15	90
Corn oil[b]	20	94
Peanut oil[a]	15	94
Inedible tallow[a]	15	88
Feed grade animal fat[c]	20	92

[a] Studies at Virginia Polytechnic Institute.
[b] Studies at University of Kentucky.
[c] Studies at Cornell University.

Root Crops

Horses in some parts of the world are fed significant amounts of
root crops, even though they are a "treat" in the United States.
Sliced carrots are routinely added to horses' rations in England
and South Africa because they are thought to help prevent hard-
working horses from going off feed. Carrots are not usually economical because they contain only about 12 percent dry matter,
and excessive feeding can restrict energy intake. Carrots are, however, an excellent source of carotene (vitamin A activity); a half-
pound of carrots per day would meet the total requirement of
vitamin A activity for a horse.

It is myth that eating large amounts of carrots will enhance
the ability to see at night. It is true that a deficiency of vitamin A
can cause night blindness, but excessive amounts of vitamin A will
not improve vision in the dark.

Turnips, sugar beets, mangels, and rutabagas have been fed
to European horses, particularly in times of war when grain was
in short supply. The dry matter content is only 9 to 11 percent,
but some of the root crops are efficiently digested. For example,
the digestibility of the organic matter of carrots is 94 percent; it is
88 percent in sugar beets.

Potatoes contain about twice the dry-matter content of the above crops, but it takes 4 to $4^1/_2$ pounds of potatoes to provide the energy provided in 1 pound of grain.

Dr. H. Meyer of Germany suggested that working horses could be fed 20 pounds of potatoes per day; mares and foals would require less. Potatoes can be fed cooked, raw, or dried.

Other Feeds

DEHYDRATED ALFALFA MEAL

This is the alfalfa plant finely ground and dried with heat. It is usually uniform in quality and composition and must meet minimum crude protein and maximum crude fiber guarantees. Pellets can be made immediately after dehydration; they are then stored in an atmosphere of inert gas, until ready for shipment, to protect the vitamin A percursors, vitamin E, and water-soluble vitamins. Dehydrated alfalfa is an excellent source of protein, minerals, and vitamins. It is often used in complete pelleted diets.

BREWERS DRIED GRAINS

This is officially defined as the dried extracted residue of barley malt, alone or in mixture with other cereal grain or grain products. It results from the manufacture of wort or beer and may contain pulverized, dried, spent hops in an evenly distributed amount not exceeding 3 percent. Brewers dried grains are high in fiber and protein. Although they have been used as a fiber source in complete pelleted rations, the digestible energy content is equal to or greater than that of oats. Wet grains that come directly from breweries should be fed fresh because of the danger of mold.

BEET PULP

Beet pulp is a by-product of the beet sugar industry and is made by drying the residual beet chips after the sugar has been extracted. Although it is often used as a fiber source, beet pulp is readily digestible; its digestible energy content about the same as oats. Beet pulp is a good source of calcium but is low in phosphorus, selenium, and vitamin A activity. It is a good fiber source in complete pelleted rations, especially for horses with heaves. Feeding moistened pulp rather than hay is also a method of reducing dust for horses with heaves.

COCOA SHELLS

Cocoa shells are the thin covering of cocoa beans. They contain about 15 percent protein, 17 percent fiber, and 10 percent ash. They are not used as a fiber source but lend flavor to livestock feeds. Cocoa shells may contain 0.5 to 2.0 percent of the alkaloid theobromine, which has been shown to stimulate appetite when fed to cattle at low levels (.05 to .1% of the diet). But caution should be used when using any cocoa by-product such as cocoa shells or cocoa bean meal in horse rations. High intakes could cause loss of appetite, diarrhea, and death. Even low levels of cocoa by-products supply enough theobromine to be detected in routine analysis of urine samples obtained from horses at races and shows. The pharmacologic effects of theobromine on horses are not well known, but preliminary studies in Ireland suggest that it may have some effect on the heart muscle and increase cardiac output. Theobromine may increase urine excretion and may also function as a smooth muscle relaxant; it has little effect on skeletal muscle.

CITRUS PULP

Citrus pulp consists of the peel and residue of the inside portion of citrus fruits used in juice production. It contains a level of acid detergent fiber almost as high as that found in hays (see Table

4.1). However, the fiber of citrus pulp is more efficiently digested than that of hay. Dr. E. A. Ott et al. (1979) at the University of Florida concluded that citrus pulp could be used as a substitute for oats without significantly changing the digestible energy content of the diet. But the crude protein content of citrus pulp is only about 7 percent and is not efficiently digested.

The biggest drawback to the feeding of citrus pulp may be that horses do not like it. Dr. Ott reported that six of eight horses given a coarse grain mixture containing 30 percent citrus pulp refused to eat it, although diets containing only 15 percent citrus pulp were readily accepted. Citrus pulp is an excellent source of calcium (1.5%) when lime is added to the pulp.

DISTILLERS DRIED GRAINS

These are obtained after the removal of ethyl alcohol from the yeast fermentation of grain or a grain mixture. Corn and rye are the grains most commonly used. Distillers grains contain 25 percent protein and 13 percent crude fiber and are useful in complete pelleted rations. The fiber can be efficiently digested so the digestible energy content is high. Distillers grains are often relished by horses.

PEANUT SKINS

Peanut skins have been used in some livestock feeds, particularly for ruminants. They have limited value for animals with simple digestive systems. When pigs were fed skins as 20 percent of the diet, their growth rate, feed efficiency, and digestibility were decreased.

We attempted to use peanut skins in pony rations, but the ponies didn't like the taste. The skins contained 15 percent protein, 33 percent NDF, 17 percent ADF, 7 percent lignin, 10 percent cellulose, and 14 percent ether extract, and 5.1 Mcal per gram. The analysis indicated that skins would be a good nitrogen source and reasonable energy source because of the high fat content. The ponies did not share our opinion, even when the skins were mixed 20:80 with a concentration of molasses and oats. So, in spite of their value, peanut skins don't appear to be popular among horses! The refusal by the ponies was probably caused by the high tannin content of the skins.

TURFGRASS CLIPPINGS

This is the material collected form turfgrass mowing operations. A. J. Turgeon and co-workers (1979) reported that the crude protein content of clippings varies widely, depending upon species and variety of plant and mowing and fertilization. Crude protein levels of 25 to 30 percent are quite common, so clippings from routine mowing may be an increasingly important feed. Dried or pelleted clippings also appear to have excellent potential as a feed-stuff, but horse owners should make sure their own lawn clippings do not contain mold or trimmings from ornamental plants that could be toxic.

WHEAT BRAN

Wheat bran is the outer covering of the wheat kernel which is removed from the cleaned and scoured wheat in the usual process of commercial milling. It contains less digestible energy but more protein, fiber, and phosphorus than whole grains. The phosphorus is not efficiently utilized because much of it is combined with the organic compound phytic acid to form a product that is not readily broken down in the horse's digestive tract.

Wheat bran is often added to horse rations because of its laxative properties. In studies at Cornell, however, the addition of 50 percent wheat bran to a hay-grain ration increased the moisture content of the feces by only a few percentage points once the horses were adjusted to the diet. Bran is often fed as a mash, but we found no evidence that steeping wheat bran improves the utilization of nutrients in the bran.

Miscellaneous Feeds

Dried bakery products are a mixture of bread, cookies, cake, crackers, flour, and dough which has been mechanically separated from non-edible material. It is a low-fiber, concentrated digestible energy source and should be used cautiously. Mold should be avoided.

Alcoholic beverages have long been traditional sources of tonics and energy. Columella (A.D. 50) advocated giving wine to horses to increase body weight. John Splan in the 1800s reported that whiskey or champagne improved appetite. The *Journal of Comparative Medicine and Surgery* claimed in July 1882 (p. 269) that thousands of racehorses have downed a pint of whiskey or brandy before running. I am not aware of any controlled studies on the benefits of alcoholic beverages for horses, but four six-packs of Black Horse ale or one quart of White Horse Whiskey could theoretically provide about one quarter of the energy required to maintain a horse.

Horses have a very high alcohol dehydrogenase activity in the liver—about 10 times as great as that in man. However, in studies at Cornell and Rutgers, horses appeared to oxidize or remove alcohol from the blood stream at a slower rate than man. So, watch your horse carefully when he noses around your liquor cabinet and don't play him slow-tempo country music. A report in the *Journal of Studies on Alcohol* indicated that people tend to drink more when listening to slow-tempo country music, and why should horses be different?

Incidentally, Sumner suggested that some human athletes perform well after moderate drinking. He reported that the winner of the 1972 Olympic marathon drank 2 quarts of beer the night before the event and that Italian athletes obtain about 4.5 percent of their calories from wine. One runner claimed to have obtained 35 percent of his calories from beer while running the Auckland-Wellington race of about 450 miles.

Many other feeds and by-products can be added to horse rations. Other examples of ingredients include grain by-products—such as wheat midds, corn bran, hominy feed, corn gluten feed, and corn gluten meal—and grape mare, pumpkins, squashes, pineapple bran, buckwheat seed, velvet beans, and yeast. Information on these and other unusual feedstuffs can be obtained from *Feeds of the World* by B. H. Schneider and *Feeds and Feeding* by F. B. Morrison.

Protein Supplements

Although protein supplements of animal origin can be used in horse rations, protein supplements of plant origin are much more commonly used. The plant proteins are usually by-products of oil extracted from seeds.

PLANT SOURCES

Soybean meal is the residue after oil has been removed from the soybean. It is the number one protein supplement for livestock in the United States. If corn is King and alfalfa is Queen of the Forages, soybean meal must be the Crown Prince of protein supplements. If the soybean hull is removed the meal contains about 49 percent protein and not more than 3 percent crude fiber. The dehulled meal is usually fed to young animals and poultry. The meal with the hulls in most horse rations contains 44 percent protein and not more than 7 percent crude fiber.

Soybean oil can be removed from the bean by mechanical extraction, or more commonly by a solvent. Raw soybean contains several heat-liable factors, such as trypsin inhibitors, which greatly reduce its nutritional value. Therefore, it is advisable to heat-treat soybean meal for horses. But don't overdo this or you'll decrease the availability of amino acids (due to the binding of some amino acids with carbohydrates). Fortunately, quality control is usually good, and soybean meal is a very consistent product.

Cottonseed meal is the second most important oilseed meal in the United States, with a production of about 8 percent that of soybean meal. Studies by G. D. Potter and co-workers at Texas A & M University have shown that cottonseed meal is deficient in lysine; however, when foals were fed lysine-fortified diets, they grew as fast and as efficiently as those foals fed soybean meal.

A second potential problem of cottonseed meal is gossypol, which has been shown to be toxic to farm animals. Pigs fed diets containing more than 10 percent cottonseed meal may lose their appetite, become weak, have difficulty breathing, and die. Congestive heart failure is common. Fortunately, horses appear to be less susceptible to gossypol poisoning than pigs.

Dr. Potter has shown that young horses can safely consume rations containing 16 percent cottonseed meal even when the meal contained 0.2 percent gossypol. Varieties of low-gossypol cottonseeds have been developed. Moise and Wysocki (1981) fed low gossypol (0.04%) cottonseed meal diets to young horses for 6 months with no apparent harmful effects, so properly supplemented cottonseed meal seems to be a reasonable protein supplement for young horses.

Peanut meal is the residue after the oil has been removed from the peanut. The amount of peanut meal in horse rations is probably quite low because the production of peanut meal is only about 0.5 percent that of soybean meal. It is usually more expensive than soybean meal and has only one half the lysine content, even though the protein content of the two meals are similar.

Linseed meal is the product remaining after the removal of most of the oil from flaxseed. The meal has long been credited with producing a "bloom" or "finish" on the hair coat, perhaps due to mucins or fats. Linseed meal extracted from flaxseed with ether contains a much lower crude fat content than when mechanically extracted. The lysine-to-protein ratio of linseed meal is much lower than that of soybean meal. It needs supplementation when fed as the primary protein source to young animals unless it is fed in high levels. In recent years soybean meal has usually been a more economical source of protein than linseed meal.

The production of linseed oil has decreased by almost two thirds during the last 30 years because the demand has greatly decreased. Vinyl floors have replaced linoleum, which linseed products were used in manufacturing. Latex paints have largely replaced linseed oil paints; other products have replaced many other uses of linseed meal.

Rapeseed meal has gained greater importance in recent years because of the tremendous increase in its production in Canada. Rapeseed is much more tolerant than soybeans of Canadian growing conditions. Rapeseed oil can be used as a salad oil, margarine oil, and shortening oil; the meal can be used in livestock rations. Rapeseed meal contains 36 percent crude protein and 2 percent lysine. The fiber content and the calcium content are about twice that of soybean meal. Early types of rapeseed meal contained high levels of goitrogenic factors (causing the thyroid gland to increase in size), but new varieties and processing methods have decreased this factor.

The protein of rapeseed was digested by horses at about the same rate as soybean meal in studies at Cornell. No problems in feed acceptability were encountered in rations containing up to 30 percent rapeseed meal.

Canadian workers found no difference in feed intake, average daily gain, or feed efficiency in growing horses fed diets containing soybean meal or diets containing 15 percent rapeseed meal.

Recently the term *canola meal* has been used for products from varieties of rapeseed containing less than 3 milligrams of glucosinotate per gram of seed.

Sunflower meal has been used mostly for feeding dairy and beef cattle because of the high fiber content (about 25 percent crude fiber). The hull can be removed to reduce the fiber content to 10 percent or less but the price is increased. Sunflower meal has a lower lysine content than soybean meal. But, if properly supplemented, sunflower meal can be fed to horses. The production of sunflowers has been spurred by the increased interest in unsaturated vegetable fats. Although still a minor crop when compared to soybeans, the production of sunflowers has greatly increased in recent years.

Alfalfa pellets, dried brewers grains, turfgrass clippings, and distillers grains all have a significant content of protein, as has been shown.

ANIMAL SOURCES

Meat and bone meal, or the dry product of mammalian tissues exclusive of hair, hoof, horn, hide trimmings, manure, and stomach contents, is used in swine and dog rations much more commonly than in horse rations. It can supply not only good-quality protein but also significant amounts of calcium and phosphorus.

Fishmeal can be made from several different kinds of fish. It is usually an excellent source of amino acids and also contains high levels of calcium and phosphorus and selenium. Although fishmeal has been fed to horses, it is seldom incorporated into horse rations because of its high cost.

Milk protein can be supplied by several products such as dried skim milk, dried buttermilk, dried whole milk, dried whey, and casein. Milk products contain an excellent combination of amino acid, plus vitamins and a good balance of calcium and phosphorus. Their great disadvantage is the price. Dried skim milk or dried buttermilk is likely to cost five to seven times more per ton than soybean meal.

Commercial Feeds

The sales of commercial horse feeds have greatly increased in the past twenty years because of their convenience. The horse owner is fortunate that many companies manufacture several different types. The horse owner doesn't have to do a great deal of calculating or buy a great variety of supplements and vitamins sources to add to the rations. The manufacturer has the equipment and facilities to utilize nutritious by-products or unusual sources of feeds; thus the commercial feed may cost less than a home mix of ingredients.

Which feed should be purchased?: The feed that does the best job, is most economical, and is provided by a company that gives service. But how can the feed be evaluated? The feed tag provides some information.

There are state and federal laws regulating the production labeling, distribution, and sale of animal feeds.

HORSE FEED

GUARANTEED ANALYSIS

Crude Protein .... (min.) 12.00%
Crude Fat .... (min.) 2.20%
Crude Fiber .... (max.) 10.00%

INGREDIENTS

Cracked Corn, Soybean Meal, Wheat Mid-
dlings, Hominy Feed, Corn Gluten
Feed, Malt Sprouts, Corn Distillers'
Dried Grains, Dried Whey, Flaked Soy-
bean Hulls, Dried Beet Pulp, Cane
Molasses, Vitamin A Palmitate (with
improved stability), D-Activated Plant
Sterol, D-Activated Animal Sterol,
Riboflavin Supplement, Niacinamide,
Calcium Pantothenate, Vitamin B12
Supplement, Vitamin E Supplement,
Menadione Sodium Bisulfite Complex,
Calcium Propionate (A Preservative),
Ground Limestone, Dicalcium Phosphate,
Iodized Salt, Traces of: Manganous
Oxide, Zinc Oxide, Iron Sulfate,
Cobalt Carbonate, Copper Oxide,
Calcium Iodate, Sodium Selenite.

GUARANTEED ANALYSIS

Crude Protein, not less than 15%
Crude Fat, not less than - 2½%
Crude Fiber, not more than 10%

INGREDIENTS

**Plant Protein Products, Grain
Products, Processed Grain By-
Products, Forage Products, Cane
Molasses, Irradiated Dried Yeast,
Vitamin A Palmitate (improved
stability), Calcium Carbonate, Di-
calcium and Monocalcium Phos-
phate, Salt, Calcium Iodate, Mag-
nesium Oxide, Manganous Oxide,
Cobalt Carbonate, Iron Carbonate,
Copper Oxide, Zinc Oxide, Potas-
sium Sulfate, Magnesium Sulfate.**

Examples of two different types of feed tags. The one on the left uses the group categories;
the one on the right uses specific feed names. (Photographs by the author)

The *Official Publication of the American Feed Control Officials* provides the following information:

1. The brand or product name must not be misleading and must be appropriate for the intended use of the feed.
2. The feed tag or label should indicate the minimum percentage of crude protein, minimum percentage of crude fat, and maximum percentage of crude fiber. Guarantees for minerals are not needed unless there are specific label claims or the feed contains more than 6 $1/2$ percent of calcium, phosphorus, sodium, and chloride. Guarantees for vitamins are not necessary unless the feed is advertised as a vitamin supplement.
3. The label should list the common or usual name of each ingredient used in the manufacture of the feed.
4. Collective terms rather than individual names can be used on feed tags. The manufacturer can take advantage of computer formulation and use alternate ingredients to provide the same levels of nutrients. The collective terms most commonly used are: Animal protein products (animal, marine, and milk products); Forage products (alfalfa meals, entire plant meals, hay, stem meals); Grain products (could be any of the grains); Plant protein products (algae, coconut, cottonseed, guar, linseed, plant, safflower, soybean, sunflower, yeast); Processed grain by-products (brans, brewers dried grains, distillers grains, distillers solubles, flour, germ meal, gluten feeds, grits, groats, hominy feeds, malt sprouts, middlings, pearled barley products, shorts, and wheat mill run); Roughage products (corn cobs, grain hulls, husks, pulps, and straws).

The feed should be appropriate for the use intended. A feed formulated for mature horses might not contain an adequate concentration of nutrients for growing horses. Or a feed for young horses may contain more protein than the mature horse needs and therefore be uneconomical. The protein source is usually one of the more expensive ingredients in the horse ration.

Feed should be purchased to complement the forage. When feeding legume hay, for example, less protein is needed in the grain mixture than when feeding grass hay.

What services does the dealer provide? Some companies offer

free forage testing or at least provide testing at a reduced rate. How reliable is the company? Is quality control good? Will the feed be available when you need it? Have other horse owners in your area had experiences—good or bad—with the feed? Availability of credit might also be a factor. Comparison shopping can often save dollars; in most areas, many options are available—so explore all of them.

PELLETED FEEDS

The percentage of commercial feeds that are pelleted has greatly increased in the last decade. There are at least four basic types of pellets. One type is the pelleted single ingredient such as dehydrated alfalfa meal. Another is the pelleted grain mixture. The third is the pelleted supplement; which may contain high levels of protein, minerals, and vitamins. The fourth type is the complete pelleted ration. This contains roughage and grain and is designed to meet all the nutrient requirements of the horse.

The complete pelleted ration appeals to many horse owners. Feed wastage is greatly reduced. On many farms, almost as much hay is wasted as is fed, but the control of waste is much easier with pellets than with hay. Storage space requirements are decreased. Costs of transportation can be reduced because pellets are less bulky than hay. Feeding of pellets can also reduce the amount of dust, which is particularly important for horses with heaves. One of the Cornell polo ponies had heaves, but after being fed a pelleted diet it was effective for several more years. Feeding a pelleted diet may reduce the appearance of the "hay belly"—an important point for show and sale yearlings.

Proper pelleting prevents the horse from sorting out the feed; thus each mouthful contains the same proportion of nutrients. Henry White of Plum Lane Farms in Lexington, Kentucky, pointed out that pelleted feeds would also reduce the possibility of feeding too many or too few supplements. Some farms have automatic feeders that dispense pellets several times during the day, decreasing labor costs and, perhaps, boredom for the horses.

Complete pelleted feeds also permit the use of feeds and by-products that would not normally be consumed by horses but have some nutritional value. The ingredient will be consumed by the horse when mixed in a pellet but might not be eaten readily if fed alone or in a simple mixture. Pelleted feeds containing cereal by-products include corn cobs, peanut hulls, high levels of fat, ryegrass straw, corn plant, pangola hay, almond hulls, tunafish meal, torula yeast, grape pulp, turkey litter, poultry litter, straw, pineapple bran, apple pomace, carrot tops, and cocoa husks.

PLAYBOY FOR HORSES

We have fed pelleted diets containing 25 percent computer paper or ground corrugated paper boxes to horses. The paper products did not contain significant amounts of protein, essential minerals, or vitamins, but they contained a high level of digestible cellulose and the energy content was simliar to that of alfalfa meal.

Of course, not just any old paper can be fed. Newspaper has a lot of ink and contains a high level of lignin, which is not digested. Glossy magazines contain a high level of clay. We fed *Playboy* magazines to ponies but found that the magazine was 30 percent dirt and indigestible!

It is likely that the use of by-products and other unusual feeds will increase as more and more horse feed manufacturers use least-cost formulations and the price of the more common ingredients increases.

But pellets have several disadvantages, too. Pelleting requires a significant amount of energy and labor and can increase the cost of feed. Unreliable manufacturers can hide poor-quality feed in pellets; you cannot look at a pellet and estimate the quality of the alfalfa, oats, or corn content. Improperly prepared pellets can become moldy because steam is added during the processing. Furthermore, it takes less time to eat pellets, so greedy eaters may be more prone to colic and digestive problems such as enterotoxemia.

Excessive heat during the pelleting process may decrease the availability of amino acids such as lysine and may destroy vitamins. I can recall one farm that had several animals with Vitamin A deficiencies because the vitamin was destroyed during pelleting.

Horses apparently feel they need a certain amount of chewing time, so include some hay in their diet or they'll start chewing their tails and wood. We found that feeding long hay in addition to the pelleted ration reduced wood chewing by ponies in stalls by 80 percent. Of course, other factors (like plain boredom) influence wood chewing too.

Horses fed pellets may have slightly softer feces than those fed conventional diets, but this should not cause many problems.

The optimum size for pellets has not been determined. Most pellets are probably $1/4$ to $1/2$ inch in diameter; however, both smaller and bigger ones are made. Henry White of Plum Lane Farm said he preferred $1/2$-inch pellets to $1/4$-inch pellets because his foals went to big pellets readily. Nagata (1970) in Japan reported that smaller pellets decreased chewing time and might also be a less satisfying source of fiber than large pellets. He also found that a greedy eater is a greedy eater, regardless of pellet size. In Nagato's studies soft pellets were preferred to extremely hard ones, but when not given a choice, the horses ate the hard ones readily.

Many materials such as sodium or calcium bentonite, lignin sulfonate, and hemicellulose extracts are added to pelleted feeds as binding agents. The type of agent used depends on the characteristics of the other ingredients; the agent is usually added at a rate of 0.5 to 2.5 percent of the ration.

The quality of the pellet can be judged by chemical analysis and visual appraisal. If overheating is suspected, the pellet can be evaluated by analyzing for fiber-bound nitrogen. The amount of "fines"—that is, fine particles that are loose and not pelleted—should be less than 5 percent.

CUBES

The commercial manufacture of hay cubes started in the 1960s. Alfalfa is the hay most commonly cubed, but other hays have been used. The cubes can be of various shapes; they are most commonly $1\frac{1}{4}$ inches by $1\frac{1}{4}$ inches by 1 to 2 inches. They can be made in the field with large self-propelled machines or from baled hay with stationary cubing plants.

Cubes have been primarily used for beef and dairy cattle; however, they can also be quite useful for horses. They have many of the advantages of pellets and more. The hay in a cube is not ground as finely as it is in a pellet; thus chewing time is increased. Horses fed cubes are also less likely to chew wood.

The analysis of cubes is often guaranteed, or at least the grade of the cube is listed. This grade depends on the maximum fiber and minimum protein contents.

Some horse owners have expressed the concern that horses fed cubes may be more prone to choke than horses fed long hay, but studies at the University of California say no. Of course, there are some choke-prone horses with a stenosis or restriction of the esophagus, and they might have more difficulty with cubes—particularly if they are greedy eaters.

Commercial feeds designed for other animals can be fed to horses under certain circumstances. They may be economical, but there are some serious dangers.

Beef cattle, poultry, or turkey rations may contain toxic levels of monensin, an antibiotic produced by *Streptomyces cinnamonensis*. The addition of monensin (rumensin) to beef cattle rations improves feed efficiency by 10 to 15 percent, but it does not increase rate of gain. Monensin changes the type of fermentation in the rumen for greater digestive efficiency, so it is very widely used in beef cattle rations.

Monensin is also widely used to control coccidiosis in poultry rations. In fact, $32 million was spent in 1978 on monensin to control poultry coccidiosis.

Although very helpful to poultry and cattle, monensin is extremely toxic to horses. The LD_{50} (dose at which it is expected that 50 percent of animals will die) is 2 to 3 mg/kg for horses but 200 mg/kg for chickens and 50 to 80mg/kg for cattle. Several cases of monensin poisoning in horses caused by the eating of poultry feed have been reported, and Whitlock et al. have noted colic, muscular weakness, lost coordination, and sweating. The horses may die within 12 to 36 hours after the first onset, due to severe damage of the heart muscles. No treatment is possible at the present time, except stall rest to relieve stress on the damaged heart.

Muylle et al. (1981) reported that they examined 32 horses with a history of poor performance several months after the ingestion of monensin sodium. Cardiac abnormalities were clearly evident in 8 horses and suspected in 4 others . They concluded that ingestion of monensin sodium by horses can cause sudden death or delayed cardiac circulatory failure.

Dairy cattle rations are often too finely ground for horses; the dust may aggravate respiratory problems and decrease intake. Dairy cattle rations containing urea should not be fed to young horses, because the urea is not efficiently utilized as a source of nitrogen.

Swine rations can also be dangerous. The antibiotic Lincomycin is often added to swine rations to increase rate of gain and feed efficiency. But Lincomycin can cause founder, diarrhea, and fatal colitis (inflammation of the colon) in horses. The Lincomycin apparently destroys certain bacteria such as those gram positive and gram negative. This allows proliferation of other bacteria, producing enterotoxins (Raisbeck and Osweiler, 1981).

Feed Preferences

Information on feed preferences of horses might be helpful in the prediction of acceptability, intake, and rate of intake of certain rations. Information should also be provided about helping entice anorexic animals to eat. Pasturing a horse will often stimulate his appetite. Of course, most trainers have their favorite methods of getting horses back on feed. John Splan, a famous harness horse driver and trainer of the 1800s, claimed that an egg in a glass of milk with a little wine stimulated his horses' appetites. If that concoction didn't work, Splan suggested a pint of champagne and half a pint of seltzer water. But we were looking for less exotic stimulants. Animals may show a preference for one of two feeds, although intake may not decrease when only the less popular feed is provided.

PASTURE

Scientists at Penn State studied the preferences of horses for bluegrass, bromegrass, timothy, alfalfa, birdsfoot trefoil, and ladino clover. Bluegrass was the first choice of the grasses; clover and alfalfa were preferred to the trefoil.

Dr. M. Archer (1978) reported that pasture with a white clover-rich mixture was more palatable than ryegrass, timothy, or orchardgrass. She compared several grasses and reported that red fescue and tall fescue were the most palatable and that perennial ryegrass and meadow foxtail were the least palatable. The grasses in the "middle" were crested dogstail, brown top, orchardgrass, and timothy.

Other studies gave the following order of pasture preference for horses; (1) red clover, (2) ladino clover, (3) smooth bromegrass, (4) meadow fescue, (5) red fescue, and (6) perennial ryegrass.

Dr. Schofield (1933) concluded that "most horses will eat other vegetation in preference to alsike clover whenever possible," and that, when consumed in large amounts, alsike clover would cause hypertrophic cirrhosis in the horse.

Preference may change during the season. Several factors—including maturity of plant and climate—may influence palatability or preference. Jordan and Martin (1975) reported that reed canarygrass can be a good pasture grass for horses early in the grazing season, but palatability decreases later on as the alkaloid content increases.

HAY

Although nutritionists in this country have generally concluded that chopping good-quality hay is not economical, Morrison (*Feeds and Feeding*) suggested that chopping poor-quality hay increases intake. Chopping is still widely practiced in many countries such as England, South Africa, and Australia. In cooperation with Dr. J. R. Gallagher of Roseworthy Agricultural College in Australia, we compared the preferences for chopped and long hay of horses accustomed to one or the other. Six horses at Cornell University that had never been fed chopped hay ate more long hay, and 8 horses at Roseworthy College that had been fed primarily chopped hay ate less long hay during a trial lasting 30 days. Thus dietary history was the deciding factor.

Other studies indicate that horses will select a red clover mixture in preference to an alfalfa mixture, assuming that the hays are of comparable quality and that timothy was preferred to orchardgrass. Horses prefer alfalfa to timothy.

GRAIN

We conducted a series of two-choice preference trials. Six ponies were given a choice of two grains for four days. The positions of the pails were alternated daily. Intakes were measured after 30 minutes at the morning feedings. The results are summarized in Table 4.11. Oats was clearly the grain preferred by the ponies. Corn was the next preferred grain, with barley third and rye and wheat on the bottom of the list. The average total intake was about 2 pounds per day, regardless of the kind of grain fed. That is, intake during the 30-minute period was not influenced by type of grain. Nutritional histories of the ponies were not available, but it is likely that the ponies had not been fed large amounts of rye, wheat, or barley prior to the trials. Thus, the preference for corn and oats may have been because the ponies were accustomed to them.

SWEET FEED

Many mammalian species have demonstrated a preference for sugars such as sucrose. It is commonly assumed that horses also have a preference for compounds that taste sweet to humans. Sugar cubes are often used for treats and the addition of molasses or sugar is frequently used by horsemen in attempts to increase feed intake. For example, one trainer reported that he bribed horses to eat by using molasses, Karo syrup, brown sugar, and honey. (It has been suggested that the increased intake that is often attrib-

Table 4.11.   Comparison of Intakes of Various Grains
by Ponies

Grains Compared	Percent of Intake	Grains Compared	Percent of Intake
Whole oats	68	Cracked corn	91
Cracked corn	32	Cracked wheat	9
Whole oats	98	Cracked corn	98
Whole barley	2	Whole rye	2
Whole oats	98	Whole barley	86
Whole rye	2	Cracked wheat	14
Cracked corn	60	Whole wheat	52
Whole barley	40	Whole rye	48
Whole corn	14		
Cracked corn	86		

uted to the use of molasses is not due to improved taste but, rather, to the decrease of dust. However, in a trial at Oregon State University horses exhibited a preference for sucrose solutions over water.)

Six ponies were used in a series of trials to study preference for sugar. The procedure was similar to the grain trials. The comparisons were oats vs. oats plus 2% sucrose, oats vs. oats plus 10% sucrose, and oats plus 2% sucrose vs. oats plus 10% sucrose.

Four of the ponies selected the diet with the greatest amount of sucrose (Table 4.12), which is not surprising. Perhaps of greater interest is the decided dislike for sucrose by one pony. Randall et al. reported that one of six horses preferred water over sugar solutions, whereas the other five preferred the sugar solutions. One pony almost always ate from the right side (92%) regardless of what he was fed; thus he averaged about 50% for all feeds.

Table 4.12. Preference of Ponies Given Choice of Oats with Levels of Sucrose

Pony No.	Oats (% of intake)	Oats + 10% Sucrose (% of intake)
1,2,3,4	18	82
5	86	14
6	46	54

Pony No.	Oats (% of intake)	Oats + 2% Sucrose (% of intake)
1,2,3,4	24	76
5	50	50
6	45	55

Pony No.	Oats + 2% Sucrose (% of intake)	Oats + 10% Sucrose (% of intake)
1,2,3,4	27	73
5	95	5
6	51	49

Five of the ponies used in the sucrose trial were also used to study the intake of oats vs. oats plus 2.5% molasses. Ponies 1,2,3, and 4 preferred oats plus molasses (63% vs. 37%), but—as in the first trial—pony 5 again preferred plain oats (74% vs. 26%) to "sweetened" oats. It can only be concluded that each horse is an individual and often displays consistently individual eating patterns.

We also compared the preferences of ponies for oats fed with or without tobacco, because several old-time horsemen claimed that horses really liked the taste of chewing tobacco. If this was so, perhaps tobacco could be used to entice a sick horse to eat. However, none of our five ponies would eat it. We tried various amounts of chewing tobacco and various kinds (Red Man, Red Horse, Work Horse, and even Camel cigarettes), but the ponies avoided the feed with tobacco. Perhaps they had read the Surgeon General's Report about the dangers of tobacco.

References

Archer, M.J. Further studies on palatability of grasses to horses. *J. Brit. Grassland Soc.* 33:239, 1978.

Easson, L. Mouldy hay. *Agr. North Ireland* 53:1, 1980.

Erickson, D.O. et al. The nutritional value of oat hay. *N. Dakota Farm. Res.* 34:13, 1976.

Fonnesbeck, P.V. Consumption and excretion of water by horses receiving all hay and hay-grain diets. *J. Anim. Sci.* 27:1350, 1968.

Goplen, B.P. Sweetclover production. *Can. Vet. J.* 21:149, 1980.

Jordan, R.M. and G.C. Marten. Effect of three pasture grasses on yearling pony weight gains and pasture carrying capacity. *J. Anim. Sci.* 40:86, 1975.

McDonald, G.K. Mouldy sweetclover poisoning in a horse. *Can. Vet. J.* 21:250, 1980.

Moise, L. and Wysocki, A. The effect of cottonseed meal on growth of young horses. *J. Anim. Sci.* 53:409, 1981.

Muylle, E. et al. Delayed monensin sodium toxicity in horses. *Eq. Vet. J.* 13:107, 1981.

Nagata, Y. Development of complete pelletized rations for racing horses. *Exp. Rep. Equine Health Lab.* 7:33, 1970.

Ott, E.A. Citrus pulp for horses. *J. Anim. Sci.* 49:983, 1979.

Potter, G.D. Cottonseed meal for horses. *Proc. Fourth Equine Nutr. Phys. Symp.* p. 19, 1975.

Raisbeck, M.F. and G.D. Osweiler. Lincomycin associated colitis in horses. *JAVMA* 179:362, 1981.

Randall, R.P. et al. Response of horses to sweet, salty, sour and bitter solutions. *J. Anim. Sci.* 47:51, 1978.

Schoeb, T.R. and R.J. Panciera. Blister beetle poisoning. *JAVMA* 173:75, 1979.

Schofield, F.W. Alsike for horses. *Ontario Vet. Coll. Reports* p. 42, 1933.

Schrug, W.A. Alternative roughage utilization by horses. *Proc. Seventh Equine Nutr. Phys. Symp.* p. 8, 1981.

Smith, K.J. Soybean meal. *Feedstuffs* Jan. 17, 1977, p. 22.

Sutton, E.I. and R.V. Stredwick. Acceptance of rapeseed meal by horses. *Can. J. Anim. Sci.* 59:819, 1979.

Swerczek, T.W. Toxicoinfectious botulism in foals. *JAVMA* 176:217, 1980.

Turgeon, A.J. et al. Crude protein levels in turfgrass clippings. *Agronomy J.* 71:229, 1979.

Waldroup, P.W. and Z.B. Johnson. Variation in protein content of soybean meal. *Feedstuffs* Sept. 20, 1976, p. 24.

Ward, G. et al. Calcium-containing crystals in alfalfa. *J. Dairy Sci.* 62:715, 1979.

Wheeler, W.A. *Forage and Pasture Crops.* New York: Van Nostrand, 1950.

Whitlock, R.H. et al. Monensin toxicosis in horses. *Proc. 24th Amer. Assoc. Equine Pract.* p. 473, 1978.

Chapter Five

Feeding Programs

If it's true that variety is the spice of life, then feeding horses can be very spicy. Every horseman has an opinion on how horses should be fed.

Naturally, not all horses should be fed in the same way, but any successful feeding program must supply the necessary ingredients: energy, protein, vitamins, and minerals. Many different feeds can be used to provide these nutrients. Of course, the feeds must be of good quality, free of mold and toxins, and the horse must like the mix.

The feeding programs outlined here are examples of how to provide the needed nutrients. Economics, availability of feedstuffs, personal preference, and many other factors may make other systems more desirable.

I agree with Professor Henry (*Feeds and Feeding*) who wrote in 1901, "The skill of the artist horse feeder enters into the very life of the creature he manages along with the food he supplies." Like people, horses are what they eat. When developing feeding programs, first determine the amount and kind of hay to be fed. The grain ration can be formulated to complement the hay, or the proper commercial grain mixture can be selected.

Unfortunately, it is often difficult to determine the amount of nutrients provided by hay. Sometimes horses are group-fed, so intake by an individual horse is not known. Even if the horses are fed individually, estimates of hay intake are often difficult to obtain, because most horse feeders do not weigh the hay and the amount of hay wasted is not known.

How Much Hay Does a Horse Need?

This question has been asked by many horsemen and nutritionists. In the late 1930s, Drs. Simms and Williams of the Storrs Experiment Station in Connecticut surveyed feeding practices of that time. More than 2,000 horses and mules used in the sugar cane, rice, and cotton fields of the South were fed an average of 25 pounds of hay per day with a range of 15 to 30 pounds. Significant amounts of that 25 pounds were wasted and trampled underfoot. Records from 40 Connecticut farms also indicated an average of 25 pounds per day for horses weighing an average of 1,350 pounds. Simms and Williams then obtained information on 50,000 city work horses in Boston, St. Louis, New York, New Orleans, Detroit, and Chicago; they concluded that the hay intake of city horses was only slightly less than that of farm horses. The doctors wanted to know whether horses could be fed less than 25 pounds. They worked with the delivery horses of the Brideport Ice Company and the R. H. Worden Dairy. Fifty-seven horses weighing an average of 1,350 pounds were fed 8, 12, or 16 pounds of hay per day for about a year. Simms and Williams discovered that 8 pounds of hay per day was sufficient.

Other scientists have suggested that the horse requires about 0.4 pound of hay per 100 pounds of body weight; thus a 1,350-pound horse would require only 5.4 pounds rather than the 8 pounds of hay suggested by Simms and Williams. We have fed lower levels successfully, but only for short periods. We fed only oats for a few months at a time. The horses did not appear to suffer at first; eventually, however, they stopped eating until they were given hay.

Of course, the horse really doesn't have to eat hay, but he appears to need a certain amount of fiber in the diet to alleviate boredom, maintain normal microbial function in the intestine, and maintain appetite. The fiber can be supplied equally well by beet pulp, citrus pulp, or any of a wide variety of feeds.

Mature draft horses at the University of Wisconsin were fed oat feed as their entire ration for 22 months. The oat feed was the

Table 5.1. Hay and Grain Ratios for Various Classes of Horse

Class	Hay %	Grain (%)
Maintenance	100	0
Mare, late gestation	65–75	25–35
Mare, early lactation	45–55	45–55
Mare, late lactation	60–70	30–40
Weanling, 6 mos.	30–35	65–70
Yearling, 12 mos.	45–55	45–55
Long yearling, 18 mos.	60–70	30–40
Light work	65–75	25–35
Moderate work	40–50	50–60
Intense work	25–35	65–75

Based on National Research Council recommendations. Hay quality will influence needed rations. Good quality hay can be fed at a higher percentage than poor quality hay.

by-product of oat grain milled for human consumption; it contained about 85 to 90 percent oat hulls with the remainder oat shorts and oat middlings. The oat feed contained about 30 percent crude fiber whereas oats contain only 11 percent. The animals ate about 48 pounds of the oat feed daily and no problems were noted. They were also given trace mineral salt. The horses did light to medium work, averaging 3 hours per day, 6 days a week; during planting, though, they often worked 11 hours per day.

A ration of only oat feed would eventually cause a deficiency of vitamin A and probably other nutrients, but the above experiment demonstrates that hay is not necessary if the ration contains other sources of fiber. It appears that the ration should provide at least 0.12 pound of fiber per 100 pounds of body weight.

There is also a maximum of how much hay or fiber should be fed. Horses with high energy requirements may not be able to meet their needs if the ration contains too much hay or fiber, because it would be bulky and have a low energy concentration. Generally, we follow the National Research Council guidelines for hay:grain ratios for various classes of horses (Table 5.1).

No one really knows how much a given horse should be fed, although National Research Council guidelines can be used as estimates. The horseman has to make the final decision based on his judgment and the horse's needs. In general, if the horse is too fat, feed intake—particularly grain—should be decreased. If the horse is too thin, feed intake should be increased. Common sense and observation will determine whether the horse is thin because of failure to utilize the feed or because of insufficient feed.

The sensitivity of the horseman to his horse's needs has been a traditional preoccupation for centuries. Perhaps the most eloquent expression was attributed to the King of Persia by Xenophon in the fourth century B.C. "It is the eye, master, which makes the horse fat."

The "eye" must consider the individuality of the horse, parasite load, environmental temperature, and quality of feeds. Other factors in working horses include the condition of the animal, skill of rider or driver, degree of fatigue, and working conditions. All factors will, of course, modify the recommendations or estimates of feed intake presented here.

The National Research Council nutrient requirements for various classes of horses are shown in Tables 5.2 through 5.7. They will be used as the basis for ration formulation. However, opinions of others will also be included.

The assumptions used in the following examples are that good-quality alfalfa hay contains 15 percent protein, 1.5 percent calcium, and .29 percent phosphorus. Mixed hay contains 11.5 percent protein, .9 percent calcium, and .23 percent phosphorus; timothy hay contains 8 percent protein, .39 percent calcium, and .23 percent phosphorus. A mixture of oats and corn is assumed to contain 10.5 percent protein, .05 percent calcium, and .4 percent phosphorus. Hay was assumed to contain 0.95 Mcal of digestible energy per pound and grain to contain 1.5 Mcal per pound. Of course, hay quality and type of grain could greatly influence these values, but the assumptions are reasonable for good feeds.

Ration formulation can be a very complicated and time-consuming activity. Owners of large numbers of horses or cattle may hire consultants or use computers to calculate rations to cut down on costs. A person with 200 horses who can save $10 per ton of grain might save as much as $16 per horse or $3200 per year.

The ration formulation system used here is very simple. The amount of nutrients provided by certain feeds will be calculated and compared first to the estimated requirements and then to the amount of supplement needed.

The nutrient requirements expressed as concentration in the total diet are shown in Tables 5.6 and 5.7. (In conversions of English to metric systems, one kilogram is equal to 2.2 pounds.)

The Mature Horse

The nutrient requirements of the mature horse are lower than for most other classes of horses. The diet should include 7.7 percent protein compared to 14.5 percent for weanlings.

The feeding program of the mature horse can be very simple. Good quality hay fed at a rate of 16 to 18 pounds of hay per 1,000 pounds of body weight, water, and trace mineral salt will provide the nutrients required for most horses at rest if there are no parasite or dental problems. Other supplements may be needed if the hay was grown on soils very low in minerals such as selenium, copper, or phosphorus.

Some grain may be needed if the horse is worked lightly, if the weather is cold, or if the horse is a hard-keeper or excessively nervous.

However, contrary to the opinion of some horsemen, a horse does not have to be fed grain. Some horses can get fat when fed only good quality hay. The grain should be regulated according to the body condition of the horse—not because the horse likes to eat oats. Of course, if hay is very expensive it may be economical to replace some of it with grain. If the cost of a ton of hay was more than two thirds the cost of a ton of grain, it makes sense to replace hay with grain because a ton of hay contains about two thirds the energy of a ton of grain.

Table 5.2. Nutrient Requirements of Horses (Daily Nutrients per Horse), Ponies, 440 Pounds Mature Weight

	Wt. (lb)	Daily Gain (lb)	Digest. Energy (Mcal)	Crude Prot. (lb)	Ca (g)	P (g)	Vitamin A Activity (1,000 IU)
Mature ponies, maintenance	440	—	8.24	0.70	9	6	5.0
Mares, last 90 days gestation	—	0.60	9.23	0.86	14	9	10.0
Lactating mare, first 3 months (12 kg milk per day)			14.58	1.56	24	16	13.0
Lactating mare, 3 months to weanling (6 kg milk per day)			12.99	1.32	20	13	11.0
Foal, 3 months of age	132	1.54	7.35	0.90	18	11	2.4
Weanling (6 months of age)	209	1.10	8.80	1.03	19	14	3.8
Yearling (12 months of age)	308	0.44	8.15	0.77	12	9	5.5
Long yearling (18 months of age)	374	0.22	8.10	0.70	11	7	6.0
Two-year-old	407	0.11	8.10	0.66	10	7	5.5

National Research Council, 1978.

Table 5.3. Nutrient Requirements of Horses (Daily Nutrients per Horse) 880 pounds, Mature Weight

	Wt. (lb)	Daily Gain (lb)	Digest. Energy (Mcal)	Crude Prot. (lb)	Ca (g)	P (g)	Vitamin A Activity (1,000 IU)
Mature horses, maintenance	880	—	13.86	1.19	18	11	10.0
Mares, last 90 days gestation	—	1.17	15.52	1.41	27	19	20.0
Lactating mare, first 3 months (12 kg milk per day)	—		23.36	2.46	40	27	22.0
Lactating mare, 3 months to weanling (8 kg milk per day)			20.20	2.00	33	22	18.0
Foal, 3 months of age	275	2.2	11.51	1.43	27	17	5.0
Weanling (6 months of age)	407	1.43	13.03	1.45	27	20	7.4
Yearling (12 months of age)	583	0.88	13.80	1.32	24	17	10.0
Long yearling (18 months of age)	726	0.55	14.36	1.30	22	15	11.5
Two-year-old	803	0.22	13.89	1.14	20	13	11.0

National Research Council, 1978.

Table 5.4. Nutrient Requirements of Horses (Daily Nutrients per Horse), 1,100 Pounds, Mature Weight

	Wt. (lb)	Daily Gain (lb)	Digest. Energy (Mcal)	Crude Prot. (lb)	Ca (g)	P (g)	Vitamin A Activity (1,000 IU)
Mature horses, maintenance	1,100	—	16.39	1.39	23	14	12.5
Mares, last 90 days gestation	—	1.21	18	1.65	34	23	25.0
Lactating mare, first 3 months (15 kg milk per day)			28.27	2.99	50	34	27.5
Lactating mare, 3 months to weanling (10 kg milk per day)			24.31	2.42	41	27	22.5
Foal (3 months of age)	341	2.64	13.66	1.65	33	20	6.2
Weanling (6 months of age)	506	1.76	15.60	1.74	34	25	9.2
Yearling (12 months of age)	715	1.21	16.81	1.67	31	22	12.0
Long yearling (18 months of age)	880	0.77	17.00	1.56	28	19	14.0
Two-year-old	990	0.33	16.45	1.39	25	17	13.0

National Research Council, 1978.

Table 5.5. Nutrient Requirements of Horses (Daily Nutrients per Horse), 1,320 pounds, Mature Weight

	Wt. (lb)	Daily Gain (lb)	Digest. Energy (Mcal)	Crude Prot. (lb)	Ca (g)	P (g)	Vitamin A Activity (1,000 IU)
Mature horses, maintenance	1,320		18.79	1.61	27	17	15.0
Mares, last 90 days gestation		1.47	21.04	1.91	40	27	30.0
Lactating mare, first 3 months (18 kg milk per day)			33.05	3.52	60	40	33.0
Lactating mare, 3 months to weanling (12 kg milk per day)			28.29	2.84	49	30	27.0
Foal (3 months of age)	374	3.08	15.05	1.85	36	23	6.8
Weanling (6 months of age)	583	1.87	16.92	1.89	37	27	10.6
Yearling (12 months of age)	847	1.32	18.85	1.98	35	25	14.0
Long yearling (18 months of age)	1,045	0.77	19.06	1.65	32	22	13.5
Two-year-old	1,188	0.44	19.26	1.63	31	20	13.0

National Research Council, 1978.

162

Table 5.6. Nutrient Concentration in Diets for Horses and Ponies Expressed on 90 Percent Dry Matter Basis

	Digest. Energy (Mcal/lb)	Crude Prot. (%)	Ca (%)	P (%)	Vitamin A Activity (IU/lb)
Mature horses and ponies at maintenance	0.9	7.7	0.27	0.18	650
Mares, last 90 days of gestation	1.0	10.0	0.45	0.30	1,400
Lactating mare, first 3 months	1.2	12.5	0.45	0.30	1,150
Lactating mare, 3 months to weanling	1.1	11.0	0.40	0.25	1,000
Creep feed	1.4	16.0	0.80	0.55	
Foal (3 months of age)	1.35	16.0	0.80	0.55	800
Weanling (6 months of age)	1.25	14.5	0.60	0.45	800
Yearling (12 months of age)	1.2	12.0	0.50	0.35	800
Long yearling (18 months of age)	1.1	10.0	0.40	0.30	800
Two-year-old (light training)	1.2	9.0	0.40	0.30	800
Mature working horse	1.0	7.7	0.27	0.18	650

National Research Council, 1978.

Table 5.7.   Estimate of Adequate
Dietary Levels of Minerals and
Vitamins

Copper	mg/kg	8
Iodine	mg/kg	0.09
Iron	mg/kg	45
Magnesium	%	0.90
Manganese	mg/kg	36
Potassium	%	0.45
Selenium	mg/kg	0.09
Sodium	%	0.30
Sulfur	%	0.13
Zinc	mg/kg	36
Vitamin D	IU/kg	250
Vitamin E	mg/kg	13
Thiamin	mg/kg	2.7
Riboflavin	mg/kg	2
Pantothenic acid	mg/kg	13

Based on National Research Council Recommendations. The ration is assumed to
contain 90 percent dry matter.

The good quality hay can be legume, grass, or a mixture. A legume would supply more protein, calcium, and vitamins than needed, but the excess would not be harmful. The legume hay would also be more likely to fatten the horse because legumes often have a greater digestible energy content than that of grass hays. But even a grass hay such as timothy could supply the horse's need.

The nutrient intake provided by 19 pounds of timothy hay and 30 grams of trace-mineralized salt is compared to the National Research Council's recommendations for a 1,100-pound horse in Table 5.8. The hay and salt provided more than NRC values for all nutrients listed except for zinc, and that requirement may have been overestimated anyway.

OPEN MARES

The nutrient requirements of the open mare are similar to those of the horse at maintenance. However, if the mare is in a

Table 5.8. Comparison of Amounts of Nutrients Provided by 18 Pounds of Timothy Hay and 30 Grams of Trace-Mineralized Salt to the Amounts of Nutrients Recommended for a 1,100-pound Horse at Maintenance

Nutrient		From Hay and Salt	Required[a] per Day	Difference
Digestible energy	(Mcal)	16.4	16.4	—
Crude protein	(lb)	1.5	1.4	+.1 lb
Calcium	(gm)	31	23	+ 8 gm
Phosporus	(gm)	17	14	+ 3 gm
Magnesium	(gm)	12	8	+ 4 gm
Zinc	(mg)	200	320	− 120[b]
Manganese	(mg)	340	320	+20 mg
Copper	(mg)	90	72	+18 mg
Iodine	(mg)	3	0.8	+ 1.2 mg
Vitamin A	(1,000 IU)	28	12.5	+13.5

[a] Based on NRC estimates of requirements.
[b] NRC requirement for zinc is probably high.

gaining state at the time of breeding, the chances for conception are increased. The practice of fattening them, or "flushing," has been used by cattlemen and swinemen, but if the mare is too fat going into breeding season, flushing may decrease chances for conception. Barren mares should be kept thin prior to breeding season.

Of course, many fat mares conceive. Are some mares barren because they are fat or fat because they are barren? If the mares do not have foals their energy requirements are greatly reduced; they are much more likely to gain weight than a mare nursing a foal. In any case, it seems logical not to keep mares too fat. Drs. Zimmerman and Green of the Ralston Purina Company reported that fat mares had a much lower conception rate than mares in good condition. On the other hand, studies at Texas A & M indicated that mares that were thin at foaling and then fed less to reduce weight during the first 90 days after foaling were unlikely to conceive again (Table 5-9). Thin mares fed to gain or maintain weight had a higher conception rate. But generally, good condition

Table 5.9.   Effects on Conception Rate of Body Condition at Foaling and Weight Changes During Lactation[a]

Body Condition at Foaling	Weight Changes[b] (lb)	Number Pregnant Number Bred[c]
Good	+ 2	7/8
Good	− 40	5/7
Thin	+ 5	5/5
Thin	− 30	1/8

[a] Adapted from data of Hennecke et al., Seventh Equine Nutrition Physiology Symposium, p. 101, 1981.
[b] From foaling weight to 90 days after foaling.
[c] At 90 days after foaling.

rather than weight seems to determine conception—in thin or fat mares.

There is little evidence to suggest that protein, vitamins, or mineral supplements will improve the conception rate if mares are fed a balanced ration. Of course, a deficiency of any of several nutrients such as protein, phosphorus, selenium, iodine, or vitamin A can reduce fertility.

THE PREGNANT MARE

The nutrient requirements of the pregnant mare early in the gestation period are not different from those of the horse at maintenance—unless, of course, the mare is also nursing a foal. Many people make the mistake of feeding the mare for two as soon as the mare is pronounced pregnant. It is only during the last 90 to 120 days of pregnancy that the fetus makes significant nutritional demands, and even then the demands are not excessive. The National Research Council suggests that the average daily energy requirement of the mare during the last third of gestation is only about 12 percent greater than for the mare at maintenance.

Nutrition of the mare during pregnancy can greatly influence the health of the foal. Dr. T. A. Mason (1981) of Australia reported that from a group of 30 pregnant mares, 17 had foals born with angular deformities of the distal radius, metacarpus, metatarsus, or tibia. These mares had been fed 25 to 30 pounds of oats per day and had access to a pasture of "dry and diminished quality" because of a drought. Small amounts of hay were given sporadically. In the following year fresh pasture was available and the mares were also given hay, oats, a calcium supplement and a trace mineral and vitamin supplement. The incidence of angular deformities was greatly decreased. Mason suggested that vitamin or mineral deficiencies may have been a problem the first year, but he thought that "overfatness" was more likely to be a problem because of the high level of oats. The mare's internal fat cramped the space for the fetus and caused congenital malformations. Later, Mason noted that in the majority of individual cases of congenital angular limb deformities presented to his clinic, the mares were overly fat.

A 1,100-pound mare at maintenance requires 16.4 Mcal of digestible energy, which is equivalent to 17 to 18 pounds of timothy hay. The National Research Council suggests that a pregnant mare of that size would need 18.4 Mcal of digestible energy. The difference of 2 Mcal can be supplied by $^2/_3$ pound of a grain mixture. As the fetus develops, the space in the mares' body cavity for the intestines and contents decreases, so hay intake should be decreased and replaced with grain. The NRC suggests that the ration of the pregnant mare be about one quarter to one third grain, depending on the quality of the hay. Thus a 1,100-pound pregnant mare would need about 13 pounds of hay and 4 pounds of grain. Estimated hay and grain intakes for mares of various weights are shown in Table 5.10.

The protein requirement for pregnant mares is about 15 to 20 percent greater than for maintenance, but the actual amount needed is not great. The 1,100-pound maintenance mare needs only 1.4 pounds of protein and the pregnant mare needs 1.65 pounds. If the 13 pounds of hay is mixed legume and grass and is assumed to contain 11 percent protein, it would provide about 1.4 pounds (13 x .11 = 1.43); if the 4 pounds of grain is a mixture of oats and the corn is assumed to contain 10.5 percent protein, it would supply .4 pound of protein (4 x .105 = .42). The total would

Table 5.10.   Estimates of Feed Intakes for Mares of Various Mature Weights

| | Weight of Mares (pounds) | | | | | | | |
| | 900 | | 1,000 | | 1,100 | | 1,200 | |
	Hay (lb)	Grain (lb)	Hay (lb)	Grain (lb)	Hay (lb)	Grain (lb)	Hay (lb)	Grain (lb)
Open	14$^1/_2$		16		17$^1/_2$		19	
Late Pregnancy	10$^1/_2$	3$^1/_2$	11$^1/_2$	4	13	4	14	4$^1/_2$
Early Lactation	10	9$^1/_2$	10$^1/_2$	10$^1/_2$	11$^1/_2$	11$^1/_2$	12$^1/_2$	12$^1/_2$
Late Lactation	11	6$^1/_2$	11$^1/_2$	7$^1/_2$	12$^1/_2$	8	13$^1/_2$	9

[a] Based on National Research Council recommendations, but actual intake needed will depend on quality of feed, type of grain, and individuality of mare.

be 1.8 pounds and a protein supplement would not be needed. If grass hay containing only 8 percent protein is fed, it would supply only 1.05 pounds of protein and a protein supplement would be needed to meet the NRC recommendation. If the protein supplement is added to the grain mixture in amounts to result in 16 percent protein, the mixture would supply 1.64 pounds of protein and the mare would be fed adequate levels. Grain mixtures calculated to contain 16 percent protein are shown in Table 5.11.

The ratio of grain-to-protein supplement to provide a required protein concentration can be calculated easily by using the Pearson Square method.

Draw a square as shown in Figures E and F. Put the protein concentration of the grain mixture near the upper left corner, the protein concentration in the protein supplement near the lower left corner, and the desired concentration in the middle of the box. Subtract the smaller number from the larger number on the diagonal and put the difference on the diagonal corner. Sum the differences and divide the number (difference) at the right top by the total of the differences. The answer times 100 is the percent

Table 5.11. Examples of Grain Mixtures Calculated to Contain Various Levels of Protein

Ingredient	12% Protein			14% Protein		
	A	B	C	A	B	C
Corn	28	46	24	41		
Oats	60	40	45	58	39	42
Barley	—	—	45	—	—	42
Soybean meal	6	8	4	12	14	10
Molasses	6	6	6	6	6	6
	16% Protein			18% Protein		
	A	B	C	A	B	C
Corn	26	39	—	25	35	—
Oats	50	35	39	45	33	36
Barley	—	—	39	—	—	36
Soybean meal	18	20	16	24	26	22
Molasses	6	6	6	6	6	6

Based on average analysis. Grain sorghum has a protein content similar to corn; barley and oats were considered to contain similar levels of protein.

of the grain. The percent of protein supplement is the bottom right number (difference) divided by the total of the differences and multiplied by 100.

Let's assume that a mixture of corn and oats containing 10.5 percent protein and soybean meal containing 44 percent protein are to be used to make a mixture of 16 percent protein. The calculations would be as shown in B of Figure E.

Of course, in these days of computers and pocket calculators, the use of equations may be preferred. Let's use the same feeds as above. Let x = percentage of grain and $100 - x$ = percentage of soybean meal. Then:

(x times concentration of protein in grain) + ($100 - x$ times concentration of protein in supplement) = $100 \times 16\%$ or:

$$(x)(10.5) + (100 - x)(44) = 100 \times 16$$
$$10.5x + 4400 - 44x = 1600$$

Change signs:
$$-10.5x - 4400 + 44x = -1600$$

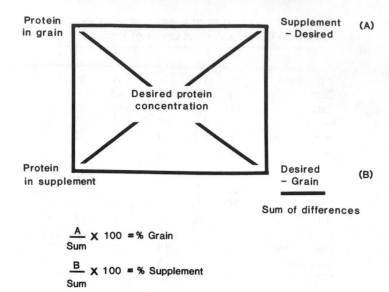

Figure E.   Pearson Square used to determine ratio of grain to protein supplement to obtain desired protein concentration in grain mixture.

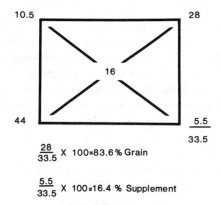

Figure F.   Example of calculations used in Pearson Square method.

Transpose:

$$44x - 10.5x = 4400 - 1600$$
$$33.5x = 2800$$
$$x = 83.6\% = \text{percentage of grain}$$
$$100 - x = 16.4\% = \text{percentage of supplement}$$

The pregnant mare must take in more calcium and phosphorus than the open mare in order to have nutrients for the bones of the fetus. The National Research Council estimates that the 1,100-pound mare will need 34 grams of calcium and 23 grams of phosphorus. If the 13 pounds of hay is legume and grass hay of average or better composition, it might provide 53 grams of calcium (13 pounds ÷ 2.2 = 5.9 kg; 5.9 kg x .9% calcium [.009] = .053 kg or 53 grams). Thus no other calcium supplement is needed. The hay would provide 14 grams of phosphorus (5.9 kg x .23% phosphorus [.0023] = .014 kg or 14 grams). The grain would provide 7.2 grams of phosphorus (4 pounds ÷ 2.2 = 1.8 kg; 1.8 kg x .4% phosphorus = .0072 kg or 7.2 grams). The total phosphorus from the hay and grain would be 21 grams (14 + 7.2), or slightly less than the estimated requirement.

If the grain mixture is supplemented with $\frac{1}{2}$-pound of a phosphorus supplement such as dicalcium phosphate (which often is the most economical inorganic source of phosphorus) per 100 pounds of grain, the horse would receive another 1.8 grams of phosphorus.

If the 13 pounds of hay is grass, it might supply only 23 grams of calcium (5.9 kg x .39 percent calcium [.0039] = .023 kg or 23 grams). The grain mixture fortified with the dicalcium phosphorus would provide about $2\frac{1}{2}$ grams of calcium. Thus an additional $8\frac{1}{2}$ grams (34 - [23 + $2\frac{1}{2}$]) would be needed. The addition of $1\frac{1}{4}$ pounds of limestone (which contains about 38 percent calcium) per 100 pounds of grain would supply 8.74 grams of calcium (.0125 x 4 pounds = .05 pounds or .05 ÷ 2.2 = .023 kg or 23 grams; 23 x .38 = 8.74 grams of calcium).

Vitamin supplements would not be needed if the hay is of good quality. If a commercial grain mixture is used, check the label for vitamins.

Selenium supplementation is recommended if the feeds were raised in areas in which the soils are low in selenium content. Trace-mineralized salt and water should be provided free choice. The mare should be given the opportunity to exercise.

Other examples of rations for pregnant mares are shown in Table 5.12. If a commercial ration is used with a legume or legume-grass hay, it should contain 10 to 12 percent protein, at least 0.1 to 0.2 percent calcium, and 0.5 percent phosphorus. Most commercial mixtures will contain calcium levels higher than 0.1 percent; levels up to 1.1 percent or perhaps higher would not cause any problems. When a commercial mixture is fed with an average grass hay, it is recommended that the mixture contain 16 percent protein, at least 0.6 percent calcium, and 0.5 percent phosphorus (Table 5.13).

Feed your mare only a light meal—perhaps a bran mash—the first day following foaling.

THE LACTATING MARE

The nutrients needed by a lactating mare depend on the amount of milk produced. Unlike dairy cattle, mares (at least in this country) have not been selected for milk production. Some mares may be heavy milkers. One report from Kentucky told of a mare that produced 50 pounds of milk daily during peak production. Reports from Europe claim that draft mares may sometimes produce more than 80 pounds of milk daily. However, this is probably rare. Some mares may not even produce enough milk to support the reasonable growth of a foal.

The National Research Council suggested that average mares may produce amounts of milk equivalent to 3 percent of their body weight daily during the first 12 weeks of lactation and an amount equivalent to 2 percent of body weight during the thirteenth to twenty-fourth week of lactation. Pony mares were estimated to produce averages of 3 and 4 percent of body weight during early and late lactation, respectively. Estimates for mares of other sizes are shown in Table 5.10. The average 1,100-pound mare in early lactation requires 28.3 Mcal of digestible energy (Table 5.4). The suggested hay-to-grain ratio is about 50:50 (Table 5.1). Thus 11 to 12 pounds of hay and 11 to 12 pounds of grain would supply the energy needs of the so-called average mare.

The 1,100-pound lactating mare requires 3 pounds of crude

Table 5.12. Examples of Grain Rations for
Mares Suggested by Various Authors

A.

Ingredient	Gestation (%)	Lactation (%)
Oats	30	15
Corn or milo	10	10
Barley	12¼	26
Wheat bran	10	7
Soybean meal	11	13
Linseed meal	4	4
Alfalfa meal	10	7
Molasses	7	7
Dicalcium phosphate	2	1¼
Limestone	¾	¾
Salt	1	1
Vitamins	2	1

B. Lactating mares (1,200 lb)
   1. Alfalfa hay (16 lb) and 6 lb of corn
   2. Red clover hay (16 lb) and 3 lb of barley, 3 lb of corn
   3. Timothy hay (16 lb) and 3 lb of oats, 3 lb of wheat bran, 1 lb of soybean meal, plus mineral supplement

C. Pregnant mares
   1. Grass-legume hay plus grain mixture containing 80% oats and 20% wheat bran, mineral mixture fed free choice
   2. Grass-legume hay plus grain mixture containing 45% barley, 45% oats, and 10% wheat bran

D. Pregnant mare (1,000 lb) 8 to 9 months
   1. Timothy hay 15 lb, corn 2 lb, oats 1 lb, wheat bran 1 lb
   2. **Pregnant mare (1,000 lb) 10 to 11 months** Alfalfa hay 14 lb, corn 3 lb, oats 3 lb, molasses 1 lb, wheat bran 1 lb
   3. **Lactating mare (1,000 lb)**
      a. Alfalfa hay 12 lb, corn 7 lb, oats 5 lb
      b. Grass hay 15 lb, corn 5 lb, oats 5 lb, molasses 1 lb, soybean meal ½ lb

Table 5.12 (continued)
E. Pregnant and lactating mares (amounts according to Table 5.10)

	Fed with Alfalfa (%)	Fed with Good Grass Hay (%)
Corn	42	36
Oats	41½	36
Soybean meal	5	15
Wheat bran	5	5
Molasses	6	6
Limestone	—	1½
Dicalcium phosphate	½	½
Vitamin supplement (if hay poor)	+	+
Trace mineral salt	fed free choice	

A. Cunha, T.J. *Horse Feeding and Nutrition*. Academic Press, 1981.
B. Morrison, F.B. *Feeds and Feeding*. Morrison Publishing Co., 1957.
C. Ensminger, E. *Horses and Horsemanship*. Interstate, 1978.
D. Bradley, M. University of Missouri, 1979.
E. Hintz, H.F. Cornell University, 1981.

protein daily. If a legume hay is used it could be expected to provide 1.8 pounds of protein (12 x .15), another 1.2 pounds of protein could come with the grain (11.5 x .105). A grass hay would provide only 0.9 pound of protein, and an additional 0.9 pound of protein would be needed. Increasing the protein concentration in the grain mixture to about 18 percent would supplement the 12 pounds of grass hay and 11½ pounds of grain mixture for the lactating mare. Once again we see the value of alfalfa hay.

If the hay was a legume-grass hay mixture, the grain mixture would need to contain about 14 percent protein if the 50:50 hay:grain ratio was used. The legume-grass hay was assumed to contain 11.5 percent protein and provide 1.38 pounds of protein (12 x .115 = 1.38 pounds of protein). Thus 1.62 pounds (3 - 1.38) would be needed in the 11.5 pounds of grain or 14 percent protein (1.62 ÷ 11.5). Examples of rations containing 14 percent protein are shown in Table 5.11. Other rations can be formulated easily by using Pearson's Square.

The 1,100-pound lactating mare in early lactation requires 50 grams of calcium and 34 grams of phosphorus. If legume hay was used (12 pounds = 5.5 kg), it would supply about 83 grams of calcium (5.5 kg x 1.5 percent calcium [.015] = 83 grams) and 13 grams of phosphorus (5.5 kg x 0.23 percent phosphorus [.0023] = 13 grams). The grain would supply 21 grams of phosphorus (5.2 kg x 0.4% [.004]= .021 kg or 21 grams). The addition of ½ pound of dicalcium phosphate per 100 pounds of grain would supply more than the needed phosphorus.

If a grass hay is fed it would supply only 20 grams of calcium (5.5 kg x .39% [.0039]= .021 kg or 21 grams). The grain mixture fortified with dicalcium phosphate would supply 10 grams of calcium. An additional 19 grams would be needed (50 - [21 + 10] = 19). If 1¼ pounds of limestone is added to 100 pounds of grain— as for the pregnant mare fed grass hay—the grain mixture would supply 23 grams of calcium (.0125 x 5.2 kg = .065 kg or 65 grams of limestone; 65 grams x .35 [limestone has 35% calcium] = 23 grams). Thus the grain mixture could be used for the pregnant and lactating mare.

Other examples of rations are shown in Table 5.12. If a commercial ration is used with a legume hay, it should contain at least 10 to 12 percent protein, 0.1 to .2 percent calcium, and 0.5 percent phosphorus. If fed with a grass hay, the grain mixture should contain 18 percent protein to meet the National Research Council recommendations (Table 5.13). The mixture should contain 0.6 percent calcium and 0.5 percent phosphorus. Although the total amounts of nutrients required for lactating mares are much greater than for pregnant mares, the concentrations of nutrients needed in the grain rations are not greatly different. The big difference is in the amounts fed and hay-to-grain ratio used.

The mare in late lactation needs slightly lower concentrations of nutrients than during early lactation (Table 5.6). However, on most farms it is easier to feed the same rations but in smaller amounts rather than formulate an additional ration. The 1,100-pound mare in late lactation requires about 24 Mcal of digestible energy. The National Research Council suggests a hay content of 60 to 70 percent and a grain content of 30 to 40 percent. The mare would thus be fed about 12½ pounds of hay and 8 pounds of grain.

Table 5.13.   Estimates of Concentrations of Nutrients Needed in Commercial Feeds for Various Classes of Horses

Nutrient		Pregnant or Lactating Mares or Yearlings		Weanlings	
		Fed with Alfalfa Hay	Fed with Grass Hay	Fed with Alfalfa Hay	Fed with Grass Hay
Protein	(%)	10–12	16–18	12–14	16–18
Calcium	(%)	0.2[a]	0.7	0.4[a]	0.9
Phosphorus	(%)	0.5	0.5	0.6	0.6

[a] Most commercial rations will contain higher levels, but this should not create a problem.

The grain mixtures formulated for the mare in early lactation would provide the necessary protein, calcium, and phosphorus.

Remember, the above calculations are only examples.

STALLIONS

Energy intake appears to be one of the most important considerations in the feeding of stallions. It makes sense not to allow the stallion to get excessively overweight; I have seen valuable animals severely foundered with their careers curtailed because of overfeeding. Experienced horsemen such as Dr. Chris Cahill (1981) of Gainesway Farms state that overweight stallions are apt to quit breeding during a heavy season. Exercise and weight control to keep the stallion in reasonable condition would minimize boredom, perhaps keep the stallion more alert and responsive, and increase the life span.

On the other hand, the stallion should not be allowed to become too thin or run-down. As a matter of fact, the ancients believed that the stallion should be on the fat side rather than on the thin side. Columella[1] (A.D. 50) said that stallions should be fattened on barley before the breeding season so they could be equal "to the fatigues of breeding." Well-fed horses in biblical times were considered to have a greater sex drive than poorly fed horses (*Jeremiah 5:8*).

The requirements for the other nutrients do not seem to be significantly greater than for maintenance. There is no reason to believe that a high level of protein would be helpful or harmful; rather, it would be a waste of money. The use of a vitamin supplement would be reasonable only if poor quality hay and grain was fed, or if a commercial feed was lacking. There is no evidence that raw eggs or wheat germ oil have a significant influence on the performance of horses fed a balanced diet.

The grain mixture described for the pregnant mare and good quality hay should make an adequate ration for the stallion. The amount of grain fed would depend on the amount of work given the stallion and the desired body condition. Some stallions may not need any additional exercise; some apparently perform better when exercised. Dr. Chris Cahill has said that exercise can be one of the most important management tools used to maintain libido.

John Williams, general manager of Spendthrift Farm, says that Seattle Slew is an easy keeper; therefore, Slew is jogged six days a week. Williams likes to keep all stallions outside as much as possible—about 18 hours per day. He feeds stallions a grain mixture of mostly oats and barley, with some corn, linseed meal, and wheat bran. He likes to feed alfalfa-timothy mixed hay. The amount given depends on the condition of the stallion.

Dr. M. Bradley of Missouri has suggested that the idle stallion may require about 15 percent more energy than a gelding at maintenance, presumably because the stallion is likely to be active. He suggested that a 1,200-pound stallion doing light breeding be fed

---

[1] But Columella also said that the stronger the stallion when it covers the mare, the greater will be the sexual vigor of its descendants. Mares can be affected by such a burning desire for breeding "that even though there is no stallion at hand, owing to their continuous and excessive passion, by imagining in their own minds the pleasures of love they can become pregnant from the wind."

10 pounds of mixed hay and a mixture containing 2 pounds of cracked corn, 4 pounds of crimped corn, 1 pound of molasses, and $1/2$ pound of soybean meal. If the mixed hay is of reasonable quality, the soybean is probably not necessary. It was suggested that the stallion during heavy breeding be fed 12 pounds of mixed hay, 9 pounds of oats, 1 pound of molasses, 1 pound of soybean meal, and 3 pounds of wheat bran.

The Young Horse

NEWBORN FOALS

The size and weight of a newborn foal can be influenced by the nutrition of the mare. Energy excess or deficiency or protein deficiency can result in smaller foals. Young mares (under 5 years of age) and old mares (over 12 years of age) are likely to have smaller foals than mares 5 to 12 years of age, possibly because they cannot provide as much nutrition to the fetus.

The size of the foal will be determined by several factors, including heritability. The average colt is bigger than the average filly. Foals born in May, June, or July may have an average birth weight 2 to 10 pounds larger than foals born in January and February.

The first milk after foaling is called *colostrum*. Adequate intake of colostrum by the foal is important for several reasons. It contains a high level of antibodies such as immunoglobulins, which combine with antigens, neutralize toxins, and agglutinate bacteria to prevent disease. In some mammals such as humans, rabbits, and guinea pigs, antibodies are transferred across the placenta, so the human baby has some protection against disease at birth. However, antibodies are not transferred across the placenta in some other mammals. The foal, calf, kid, and lamb are born without significant levels of immunoglobulins and should therefore receive colostrum. They do not develop their own defense system until two to three months after birth.

The failure to acquire immunoglobulins is often stated to be the single most important factor causing infection and death in a significant number of newborn foals. A survey of 87 Thoroughbred foals in Washington showed that 24 percent had lower than desired levels of immunoglobulins.

Low levels of immunoglobulins due to lack of colostrum can be caused by the death of the mare at foaling, by failure of the mare to produce colostrum (which is most common in mares foaling for the first time), or rejection of the foal by the mare. Dr. K. Houpt of Cornell University suggests that the chances for mare rejection of the foal can be decreased by reducing disruption in the foaling stall during the first one or two hours after foaling. Let the mare and foal get acquainted. Don't rush to give an enema. Don't immediately invite all the friends and family in to see the new champion. Don't be in a hurry to clean the stalls; odors are important in developing a mare-foal bond.

Low immunoglobulin levels are often caused by colostrum leaking from the udder before the mare foals. This can be triggered by infection of the placenta, but sometimes there is no apparent reason for it.

The foal should receive 1 to 2 pints of colostrum within 24 hours after birth. After that time the digestive tract changes and the immunoglobulins cannot be absorbed.

Many farm managers milk some colostrum from normal mares and freeze it for later use. Dr. Leo Jeffcott (1974) at the Equine Research Center in Newmarket, England, concluded that it is usually safe to collect 6 to 8 fluid ounces from a normal healthy mare without depriving her foal. He suggested that the mare's foal be allowed to suck first; then the colostrum can be milked into a clean plastic container. The colostrum will last for a year if stored in a freezer at -15 to -20°C. Thawing and refreezing could destroy the antibody activity. Jeffcott suggested that foals deprived of their mare's colostrum should be given not less than 3 bottles, each containing 6 to 8 fluid ounces, by bottle feeding or stomach tube feeding as soon after birth as possible. He suggested that the colostrum be given hourly in three feedings.

If frozen colostrum is not available, deprived foals can be given plasma from male horses that have never received blood transfu-

sions. Mares that have foaled or males that have received trans-
fusions may have antibodies capable of destroying the red blood
cells of the foal.

Freeze-dried serum and freeze-dried colostrum have also been
used successfully but are not commercially available at this time.

Colostrum provides more than antibodies. It has about five
times the protein concentration and twice the energy concentration
found in milk. Colostrum is also an excellent source of vitamin A.
Studies have suggested that foals may be born with low levels of
vitamin A and, when deprived of colostrum, have much lower
serum levels of vitamin A than other foals. The vitamin A requirement for a newborn foal might be 3,000 to 4,000 IU daily; therefore, foals deprived of colostrum might benefit from vitamin A
supplementation.

Many young foals have been observed eating their dam's feces.
Owners often become concerned because they feel the foal must
be lacking something. The eating of feces (coprophagy) is usually
most common during the second to fourth weeks after foaling.
Frances-Smith and Wood-Gush (1977) reported that coprophagy
was a normal part of a foal's development, as it introduced bacteria
into the foal's gut and supplied protein, vitamins, and minerals.
However, coprophagy also infests the foals with parasites.

ORPHAN FOALS

Nurse mares (if available) are a very convenient method of
rearing orphan foals. Some mares will readily accept foals; others
may require tranquilizers.

If nurse mares are not available, a nurse goat of one of the
dairy breeds such as Toggenburg, Saanen, or French Alpine can
be used. Nubian goats can also be used, but they give less milk than
the other breeds and the milk is richer in fat. Many nannies readily
accept foals. The foal can be trained to nurse on its knees; however,
it is better to put the goat on a platform so that her udder would
be about the same height above the ground as the mare's udder.
In some areas of this country farms have nurse mares or nurse
nannies available for lease. Goat's milk usually has a higher con-

centration of protein and fat and a lower level of lactose than mares' milk, but apparently the differences are not so great as to cause problems for the foal.

Cow's milk is much richer in fat and protein than mare's milk. It should be diluted before being given to foals because if not it may cause diarrhea. Cows' milk is lower in lactose than mare's milk. F. B. Morrison (*Feeds and Feeding*) suggested that 4 ounces of lime-water and 1 teaspoonful of sugar be added to 16 ounces of cow's milk to make a suitable mixture for foals. (Limewater is made by adding an excess of calcium oxide or calcium hydroxide to water and leaving it well stoppered over night. The clear solution can be poured off the next morning. If there is no residue, not enough calcium hydroxide was added.) Morrison further suggested that the foal be fed about 8 ounces of the mixture every hour for the first day or so, which would provide a total of 192 ounces (6 quarts). By the third day the foal could be fed 6 quarts in four equally spaced feedings. He said that after a few more days the foal could be fed unmodified whole milk.

We have found commercial milk replacers such as Borden's Foal Lac® to be very useful. The average body weight of four-month-old foals weaned at birth and fed Foal-Lac was similar to the average body weight of foals left with their mothers. We found the feeding regimen recommended by Bordens to be reasonable, but the concentration of Foal-Lac and intake of some foals had to be adjusted by trial and error to prevent constipation or diarrhea.

SUCKLING FOALS

The mare reaches her peak production about three to four months after foaling, and production usually decreases rapidly after that. The concentration of nutrients in mare's milk also decreases during the lactation period. Studies at Michigan State University indicated that the crude protein content of mare's milk the first few days after foaling was 3 to 4 gm per 100 gm of milk, but by four months this dropped to only 2 gm per 100 gm. The concentration of fat decreased by 50 percent during the same period.

Mares are not selected for their ability to milk, so many are poor milk producers. Creep feeding—in an area where the foal can eat grain without being hindered by the mare—allows faster growth and may help prevent setbacks at weaning. Some authorities, however, have suggested that creep feeding may promote such rapid growth that skeletal problems such as epiphysitis can develop. Therefore, never force-feed or stimulate intake with hormones.

The National Research Council recommends that creep feed contain 16 percent protein, 0.8 percent calcium, and 0.55 percent phosphorus (Table 5.6). Grain should be cracked, crimped, or rolled for young foals. Examples of creep rations are shown in Table 5.14.

WEANLINGS

Rations required for weanlings depend on the age at which the foals are weaned. The nutrient concentrations required in the ration of a three-month old foal is much different from that of a six-month old foal (Table 5.6).

The amount and concentration also depends on the desired rate of gain, which is much faster now than in the past (Table 5.15). For example, a horse with an average mature weight of 1,100 pounds might easily be expected to gain $2\frac{1}{4}$ pounds per day during the first 6 months of life and to weigh 500 pounds at 6 months of age. Earlier estimates would suggest $1\frac{3}{4}$ pounds per day, with a body weight of 440 pounds at 6 months of age. Many horse owners want the foal to grow even more rapidly than the values in Table 5.15 suggest in order to be ready for yearling sales and shows.

Height increases faster than body weight (Table 5.16). Proper nutrition during the period of rapid bone elongation is essential to develop a sound skeleton; the increasing pressure on rapid rate of gain may be responsible for the increased incidence of skeletal problems in young foals.

Experiments have demonstrated that overfeeding and rapid growth cause skeletal diseases in dogs, cranes, pigs, and cattle. Overfeeding and rapid growth have long been considered by many to be harmful to horses. This is not simply a modern phenomenon. As far back as 1856, G. H. Dadd wrote, "Many thousands of horses

Table 5.14.   Examples of Creep Rations

A.	Ingredients (%)	Fed with Alfalfa Hay
	Cracked corn	38
	Crushed oats	37
	Molasses	6
	Soybean meal	17
	Limestone	1
	Dicalcium phosphate	1

B.	Ingredients (%)	Fed with Mixed Hay
	Cracked corn	53
	Soybean meal	33
	Molasses	10
	Trace mineral salt	1
	Limestone	1
	Dicalcium phosphate	1
	Brewers' yeast	0.5
	Vitamin supplement	0.5

C.	Ingredients (%)	Fed with Alfalfa Hay
	Cracked corn	30.8
	Crushed oats	61.5
	Soybean meal	7.7
	Minerals added as needed	

A. H.F. Hintz, Cornell University.
B. William Tyznik, Ohio State University.
C. M. Bradley, University of Missouri.

die in consequence of being too well or rather injudiciously fed. Men who prepare horses for the market attempt to get them into condition without regard to their general health." Henry (*Feeds and Feeding*) in 1901 wrote, "In no other way can colts be ruined so surely and so permanently as by liberal feeding and close confinement."

Table 5.15.   Percent of Mature Weight Attained at Various Ages

| Breed | Year | Reference | Age (months) | | |
			6	12	18
Standardbreds	1905	1	36	58	70
Morgans	1945	2	40	61	74
Morgans	1923	3	42	61	75
Grade	1945	2	41	63	74
NRC	1949	4	40	58	70
	AVERAGE		39.8	60.0	72.6
Thoroughbreds	1969	5	48	68	82
Standardbreds	1974	6	44	64	79
Arabians	1977	7	46	66	80
Quarter Horse	1961	8	44	63	79
Anglo-Arabs	1971	9	45	67	81
Thoroughbreds	1979	10	46	67	80
NRC (400 kg)	1978	11	46	66	83
NRC (500 kg)	1978	11	46	65	80
Morgans	1981	12	46	65	81
Standardbreds	1981	13	45	67	—
	AVERAGE		45.7	65.9	80.6

1. Henry, *Feeds and Feeding*, 1912.
2. Dawson, W.M. et al., *J. Anim. Sci.* 4:47, 1945.
3. University of Vermont. *Exp. Stat.*, 1923.
4. National Research Council. *Nutrient Requirements of Horses*, 1949.
5. Green, D.A. *Brit. Vet. J.* 124:539, 1969.
6. Hintz, H.F. Unpublished data.
7. Reed, R.R. and N.K. Dunn. *Fifth Equine Nutr. Phys. Symp.* p. 99, 1977.
8. Cunningham, K. and S. Fowler, *La. State Exp. Sta. Bull.* 546, 1961.
9. Budzynski, M.E. et al. *Recz. Nauk. Roln. Series B* 93:21, 1971.
10. Hintz, H.F. et al. *J. Anim. Sci.* 48:480, 1979.
11. National Research Council, *Nutrient Requirements of Horses*, 1978.
12. Balch, D. University of Vermont, personal communication, 1981.
13. Person, B. and R. Ullberg. *Eq. Vet. J.* 13:254, 1981.

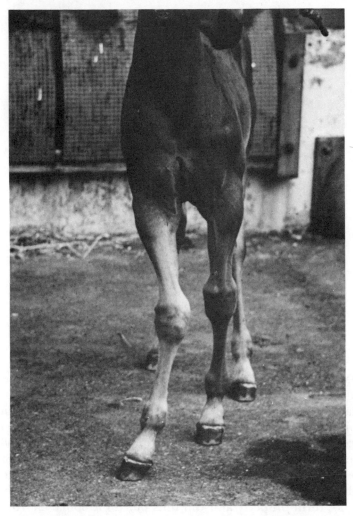

Angular deformity, commonly called knock-knees, caused by unequal growth in growth plate. Extent of nutritional involvement is unknown; can often be corrected surgically. (Courtesy K. K. White, Cornell University)

Table 5.16.   Percentage of Mature Height at
Various Ages

Breed	Reference	Age (months) 6	12	18
Thoroughbred	1	83	90	95
Arabians	2	84	91	95
Anglo-Arabs	3	83	92	95
Quarter Horses	4	83	91	96

1. Hintz et al. *J. Animal Sci.* 48:480, 1979.
2. Reed and Dunn. *5th Equine Nutr. Physiol. Symp.* 1977.
3. Buydzynski et al. *Recz. Nauk. Roln. Series B* 93:21, 1971.
4. Cunningham and Fowler. *La. Agr. Exp. Stat. Bull.* 546, 1961.

OTHER SKELETAL PROBLEMS

Many of the skeletal problems of young horses are most commonly found in rapidly growing, heavily muscled animals—particularly those with fine bone. Stromberg (1979), a Swedish scientist, reported that feeding high amounts of grains increased the incidence of osteochondrosis in young horses (a condition in which pieces of cartilage in the joints of bones become detached from the bone). Overfeeding has also been suggested as a possible factor in producing wobbles. The spinal column in the rapidly growing horse may not form properly and can thus pinch the spinal chord.

Dr. D. Kronfeld (1978) suggested that overfeeding of genetically predisposed animals might be responsible for skeletal disorders like epiphysitis, which is characterized by abnormalities of the long bones such as the radius (main bone of forearm), metacarpus (front cannon), tibia (main bone of the gaskin), and metatatarsus (rear cannon). The physes—or growth plate—may be enlarged and irregular. The condition can be found in animals that are fed inadequate levels of calcium, but it can also be found in animals apparently fed a balanced diet. In the latter case the animals are usually growing very rapidly.

"Contracted tendons" causes a decrease in the angle of the fetlocks and pastern joint. It is unlikely that the tendon actually contracts, but the relative growth or development of the bone, muscle, or tendon is somehow out of balance.

Overfeeding has been incriminated in the development of contracted tendons. Underfeeding followed by overfeeding may also result in contracted tendons. We restricted the average weight gains of foals weaned at four months to about $1/2$ pound per day, by limiting feed for four months. The foals were then fed the same diet free choice. Another group of foals were fed free choice from weaning. They gained almost two pounds per day during the four months. When the restricted foals were fed free choice they grew rapidly and four of six foals developed contracted tendons within two months. No problems were observed in the foals fed free choice.

Unfortunately, the mechanisms by which rapid growth and/or overfeeding are related to skeletal problems are not clearly defined, but genetics could be involved.

Young horses fed high levels of energy and protein grow rapidly and are therefore more susceptible to deficiencies of other nutrients. A foal that is growing rapidly needs more grams of calcium per day to form sound bone than a foal growing at a slow rate. But a high protein level per se does not seem to be a problem. We have fed protein levels as high as 24 percent without inducing skeletal problems or interfering with mineral metabolism when the diet was balanced.

So, balance the diet carefully to include the desired levels of energy, protein, minerals, and vitamins. Then feed your horses according to the desired rate of gain, but watch the animals closely! If problems such as epiphysitis or contracted tendons start to develop, feed intake should be decreased. Many affected animals apparently respond to such a treatment and recover.

Growth Rates for Young Horses

The optimal growth rate is not known, but estimates of body weights of light horses at various ages are shown in Table 5.17. Remember, these are only guides; the foals must be treated as individuals.

Table 5.17.   Estimates of Body Weight
at Various Ages for Light Horses
of Various Mature Body Weights

	Mature Weight (lb)		
Age	900	1100	1300
MONTHS			
2	230	290	330
4	325	400	465
6	415	510	585
8	485	600	700
10	545	680	790
12	600	740	850
14	640	790	900
16	680	840	950
18	710	880	1000

Few recent trials have been conducted with draft horses. The data in Table 5.18 were taken from several early experiment station reports. Nevertheless, they suggest that heavy horses may attain mature weight at a slower rate than light horses. The average percent of mature weight at 6, 12, and 18 months was 37, 56, and 70 percent respectively, compared to 46, 67, and 80 percent for

Table 5.18.   Body Weight of Draft Horses at Various Ages

Source	Year	Age of Horse (months)					
		6	12	18	24	48	60
Iowa	1943	666	1,105	1,245	1,505	1,700	1,800
Canada	1934	730	1,020	1,350	1,480	1,790	2,050
Missouri	1933	542	876	1,150	1,250	—	1,500
Montana	1945	610	960	1,100	1,235	1,400	1,520
Cornell	1921	544	832	1,104	—	—	1,600
AVERAGE		618	959	1,190			1,694

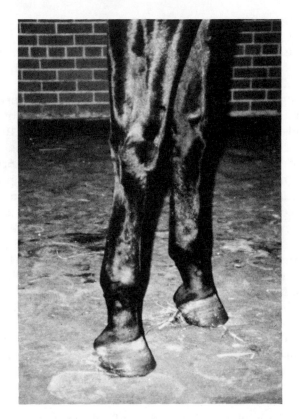

A metacarpal-phlangeal flexure deformity commonly called "contracted tendons." Most commonly seen in rapidly growing horses. (Photograph by the author)

light horses. Bulletins of the 1920s to 1940s recommended that draft foals make about one half of their mature size during the first year and three quarters by two years of age.

Let's assume that the example foal has an expected mature weight of 1,100 pounds. According to the National Research Council, at 6 months the foal requires 15.6 Mcal of digestible energy and can be fed a hay-grain ration of 65 to 70 percent grain and 30 to 35 percent hay. A ration of $8\frac{1}{2}$ pounds of grain and $3\frac{1}{2}$

pounds of good hay would provide the needed energy. The foal needs 1.74 pounds of protein. If the hay is alfalfa, it might provide .52 pounds of protein (3.5 x .15); thus 1.22 pounds (1.74 − .52) is needed from the grain. The grain should contain 14.3 percent protein (1.22 ÷ 8.5).

Examples of grain mixtures containing 14 percent protein are shown in Table 5.11. If the hay is timothy it might provide only .28 pounds of protein (3.5 x .08) and an additional 1.46 pounds (1.74 − .28) would be needed. A grain mixture containing 17 percent protein (1.46 ÷ 8.5) would supply the needed protein. If the hay is mixed alfalfa and grass, it would provide .4 pounds of protein (3.5 x .115) and 1.34 (1.74 − .4) would be needed. Grain mixtures containing 16 percent protein are shown in Table 5.11.

The example foal needs 34 grams of calcium and 25 grams of phosphorus. The grain mixture might supply 15.6 grams of phosphorus (8.5 pounds ÷ 2.2 = 3.9 kg; 3.9 kg x .004 = 0.156 kg or 15.6 grams); the hay might supply 3.7 grams of phosphorus (3.5 pounds ÷ 2.2 = 1.6 kg; 1.6 kg x .0023 = .0037 kg or 3.7 grams). The total grams of phosphorus would be 15.6 + 3.7 or 19.3, and so 25 − 19.3 = 5.7 grams more would be needed. The addition of 1 pound of dicalcium phosphate per 100 pounds of grain would provide an additional 7.2 grams of phosphorus and meet the requirements. One pound of dicalcium phosphate per 100 pounds of grain would provide slightly more phosphorus than needed.

Alfalfa might supply 24 grams of calcium (1.6 kg x .015 = .024 kg or 24 grams). The grain supplemented with dicalcium phosphate would provide 10 grams (3.9 kg x .0026 = .001 kg or 10 grams); thus the requirements of 34 grams of calcium would be met.

Mixed legume and grass might provide 14.4 grams of calcium (1.6 kg x .009 = .0144 kg or 14.4 grams). The hay and grain would provide only 24.4 grams (14.4 + 10), so 9.6 grams would be needed.

If one pound of limestone was added per 100 pounds of grain, it would provide 13.7 grams of calcium (.01 x 3.9 = .039 kg or 39 grams; 39 grams x .35 [as limestone is 35 percent calcium] = 13.7 grams).

Timothy might supply 6.4 grams of calcium (1.6 kg x .004 =

.0064 kg or 6.4 grams), and then only 16.4 grams (6.4 + 10) would be provided; 17.6 (34 - 16.4) more grams would be needed. The addition of $1\frac{1}{2}$ pounds of limestone per 100 pounds of grain would provide an additional 21 grams of calcium (0.015 x 3.9 kg = .059 kg or 59 grams; 59 grams x .35 = 21).

Some horsemen become concerned that if their foals do not grow rapidly they will not attain their potential size. However, several experiments have dispelled these fears if growth is not greatly retarded.

Professor Hudson (1933) fed two groups of draft colts 13 pounds of grain and 10 pounds of hay, or 3 pounds of grain and 8 pounds of hay, daily for the first 4 months following weaning. The liberally fed colts gained 292 pounds and grew 4.2 inches taller. The foals on limited feed gained 144 pounds but only grew 3.2 inches taller. During a second period of 5 months, the liberally fed foals were given $8\frac{1}{4}$ pounds of grain, 2 pounds of hay and pasture; the limited-feed foals were given $\frac{2}{3}$ pound of grain, 2 pounds of hay and pasture. The liberally fed foals gained another 253 pounds and grew another 3 inches, but the limited-feed foals gained 280 pounds and grew another 2.7 inches. Apparently, the limited fed foals ate *more* pasture than the liberally fed foals. The average total gains after 9 months were 545 pounds for the liberal group and 427 pounds for the limited group.

Hudson concluded that although liberal feeding hastens maturity and keeps colts in better sales condition, the ultimate body development is not all that different, because the average mature body sizes of both groups were similar.

There is no doubt that early and prolonged limited feeding can decrease mature size. Nevertheless, the horse has a remarkable ability to recover. Of course, if the foal has been limited he should be watched closely when given liberal amounts of grain. If skeletal problems develop, decrease grain intake.

Examples of diets for weanlings are shown in Table 5.19. Estimates of feed intake based on National Research Council recommendations are shown in Table 5.20. Estimates of concentrations needed in commercial rations are shown in Table 5.13.

Many factors other than diet can influence rate of gain. Disease

Table 5.19. Examples of Rations for Weanlings

Ingredient	%	Ingredient	%	Ingredient	%	Ingredient	%
A		**B**		**C**		**D**	
Corn, barley, or combination	30¾	Cracked corn	36	Cracked corn	43	Cracked corn	30.5
Oats	25	Oats	30	Oats	38	Oats	61.5
Milo (or corn or barley)	15	Soybean meal	25	Soybean meal	11	Soybean meal	8.0
Soybean meal	15	Molasses	5	Molasses	5		
Alfalfa meal	5	Limestone	2	Limestone	1		
Molasses	5	Dicalcium phosphate	1	Dicalcium phosphate	1		
Dicalcium phosphate	2	Trace mineral salt	1	Trace mineral salt	1		
Trace mineral salt	1						
Vitamin supplement	¾						
Limestone, ground	½						

A. Cunha, T. *Horse Feeding and Nutrition.* Academic Press, 1981. Ration designed to be fed with hay containing at least 12% protein.

B. Cornell University. Feed with good quality grass hay.

C. Cornell University. Feed with good quality legume hay.

D. Bradley, M. University of Missouri. Fed with good mixed hay. Minerals added according to mineral content of hay.

Table 5.20.   Estimates of Feed Intakes for Growing Horses of Various
Expected Mature Weights

Age (months)	Expected Mature Weights (pounds)							
	900		1000		1100		1200	
	Hay	Grain	Hay	Grain	Hay	Grain	Hay	Grain
	(lb/day)							
3	$1\frac{1}{4}$	7	$1\frac{3}{4}$	$7\frac{1}{2}$	$2\frac{1}{2}$	8	3	$8\frac{1}{2}$
6	2	$7\frac{1}{2}$	$2\frac{3}{4}$	8	$3\frac{1}{2}$	$8\frac{1}{2}$	4	$8\frac{3}{4}$
9	5	6	$5\frac{1}{2}$	$6\frac{1}{2}$	6	7	$6\frac{1}{2}$	$7\frac{1}{2}$
12	$5\frac{3}{4}$	$5\frac{3}{4}$	$6\frac{1}{2}$	$6\frac{1}{2}$	$7\frac{1}{2}$	7	$7\frac{3}{4}$	$7\frac{1}{2}$
15	$6\frac{1}{2}$	$5\frac{1}{4}$	7	6	8	$6\frac{1}{2}$	$8\frac{1}{2}$	7
18	8	$4\frac{1}{2}$	$8\frac{1}{2}$	5	9	6	$9\frac{1}{2}$	$6\frac{1}{2}$

Based on National Research Council estimates of energy requirements. These
estimates may be useful as guidelines. Actual intakes of feed will depend on
quality and kind of feed, individuality of foal, and desired rate of gain.

and parasites can greatly decrease development. Climate may also
have an effect. Colts may grow slightly faster than fillies. Some
hormones and anabolic steroids may increase rate of gain, but they
are not recommended at present because their long-term effect on
the foal is not known. Glucocorticoids may decrease rate of gain
and development of long bones.

Antibiotics have been used to stimulate growth rate in poultry
and swine but are not widely used in horse rations. Chlortetracy-
cline has been used, but monensin and lincomycin can be toxic.

Exercise is important. It firms muscles and stimulates appetite.
However, heavy forced exercise for a youngster may cause problems.

Raising foals in a three-sided run-in shed allows exercise, saves
labor, ensures adequate ventilation, and therefore decreases the
risk of respiratory problems. A complete feed—either pelleted or
a mixture of chopped hay and grain—works well with a run-in
shed. When the foals become accustomed to the ration and there
is adequate space per foal so the boss foal doesn't eat more than
his share, the feed can be fed free choice without great danger of
overfeeding.

Don't let the feed become stale or moldy. When free-choice feeding, always make feed available. Don't allow foals to run out of feed on week-ends and then give unlimited feed on Monday; this can cause severe digestive upset, colic, enterotoxemia, or founder. Grain should never be fed free choice by itself because of the danger of overfeeding; always mix it with hay in a pellet or with chopped hay. The ratio of hay to grain can be increased with time because the rate of gain decreases with time. If the hay-to-grain ratio is not changed, the foals may become too fat. If a complete pellet is used include some hay to help prevent wood chewing.

THE YEARLING

The rate of gain is much slower for yearlings than for weanlings. Therefore, the concentrations of energy, protein, minerals, and vitamins required by yearlings are less than those required by weanlings—although total intake may be greater. The nutrient requirements of the yearling are not greatly different from those of the pregnant or lactating mare. So, simply follow the mare's ration. Estimates of intake are shown in Table 5.20.

Performance Horses

RACEHORSES

Few controlled experiments have been conducted in the nutrient requirements of racehorses. The National Research Council's recommendations for energy were based on limited studies with horses performing other activities (Table 5.21) and were applied to racehorses. According to the National Research Council, the energy requirements for a Thoroughbred or Standardbred at the racetrack would be 20 to 25 Mcal per day per 1,000 pounds of body weight. But the results of surveys conducted at four Stan-

Table 5.21. Energy Requirements for Various Activities
of Horses[a]

Activity	DE/hr/kg of Body Weight[b] (kcal)
Walking	0.5
Slow trotting, some cantering	5.0
Fast trotting, cantering, some jumping	12.5
Cantering, galloping, jumping	23.0
Strenuous effort (polo, racing at full speed)	39.0

[a] National Research Council, 1978.
[b] Above maintenance requirement.

dardbred and two Thoroughbred racetracks indicate that race-horses need to be fed much more than these estimates.

L. D. Winter (1980) interviewed 7 trainers with 55 horses at Finger Lakes Racetrack and 8 trainers with 50 horses at Belmont Racetrack in the fall of 1979. The average weight of the horses at both tracks was 1,075 pounds. Grass hay and oats were the basis of all diets. The horses at Finger Lakes were fed $14\frac{1}{2}$ pounds of hay (mostly timothy), $13\frac{1}{2}$ pounds of oats, and 2 pounds of sweet feed. Three trainers fed about 1 pound of wheat bran once a week, two trainers replaced $1\frac{1}{2}$ pounds of oats with corn, and two others replaced 1 pound of oats with barley. The horses at Belmont were fed an average of 16 pounds of hay, $12\frac{3}{4}$ pounds of oats, and $2\frac{1}{2}$ pounds of sweet feed. Five of the six trainers fed 1 pound of wheat bran (usually in a mash) once a week. One of the eight replaced $1\frac{1}{2}$ pounds of oats with corn. The average daily intake of digestible energy at Finger Lakes was 35.8 Mcal; that at Belmont was 35.6 Mcal, or about 10 Mcal greater than estimated by the National Research Council.

In other studies, a total of 50 trainers of Standardbreds were interviewed at four tracks (Vernon Downs, Roosevelt Raceway, Meadowlands, and Pompano Park). The trainers were responsible for 232 horses. Oats were the primary grain for all but one trainer (who fed corn with fewer oats and less sweet feed than the others). He fed it year round and reported no problem, even though others modified their corn feeds because they felt it was "hot food." The average total grain intake was 1.5 pound per 100 pounds of body weight.

Grass hays (usually timothy) or a mixed hay of mainly timothy were the primary roughages at the northern tracks, although some alfalfa was fed. Alfalfa cubes were popular at Pompano Park (Miami, Florida) because trainers there found it difficult to get high quality hay. Four of the eight trainers interviewed at Pompano said they fed alfalfa cubes as the only roughage source and reported no difficulties. The other four trainers fed some hay with the cubes. One trainer reported that some horses had problems with irritated gums unless the cubes were softened by soaking in water. The average roughage intake was 1.8 pounds per 100 pounds of body weight. The results are summarized in Table 5.22. The energy intake was again greatly higher than recommended by the National Research Council, but the horses were not gaining weight.

Del Miller reported that a Standardbred would eat 15 to 20 pounds of hay and 12 to 15 pounds of grain per day, which agrees with estimates obtained in the above estimate. Dr. William Tyznik suggested that the total intake would be about 3 pounds per 100 pounds of body weight—which again is not greatly different from the above survey estimates, but much more than the National Research Council estimates.

The trainers in the above survey usually changed their feeding programs on race days. About half fed no hay before the race; the other half fed greatly reduced amounts.

The protein requirement is not greatly increased during racing since muscle is not used up. The National Research Council suggests that a 1,100-pound horse at maintenance requires 1.4 pounds of protein daily (or about 8 percent protein).

Good quality oats contain 12 percent protein or more, corn 9 percent or more, and even good quality grass hay should contain more than 8 percent. Thus a ration with good quality feeds including oats would usually provide 10 percent or more protein. Some nitrogen is lost in the sweat, but it will be replaced if the

Table 5.22. Feed Intakes at Four Standardbred Racetracks and Estimates of Energy Intake[a]

Grain Fed	Number of Trainers	Hay or Cubes	Oats	Corn	Sweet Feed	Digestible Energy
		(lb/day)	(lb/day)	(lb/day)	(lb/day)	(Mcal/1,000 lb body wt.)
ROOSEVELT RACEWAY						
Oats	12	19[b]	10	—	4½	36.0
Oats and Corn	6	20	9½	1½	3½	36.8
VERNON DOWNS						
Oats	11	16	10½	—	4	34.7
Oats and Corn	4	17	8½	2	3	34.5
MEADOWLANDS						
Oats	4	21	12	—	5	42.6
Oats and Corn	5	20	8	6	3	42.9
POMPANO PARK						
Oats[c]	8	16[d]	8	—	7	37.2

[a] Nash, R. and H.F. Hintz. *Hoofbeats*. Sept. 1981, p. 27.
[b] One trainer in each group fed alfalfa cubes as the primary roughage.
[c] No corn was fed by trainers interviewed at Pompano.
[d] Four trainers fed only alfalfa cubes (av. 15 lb/day).
Two trainers fed only alfalfa hay (17 lb/day).
Two trainers fed alfalfa hay (10 lb) and alfalfa cubes (7 lb/day).

ration is adequate when the horse eats more feed for energy.

The average calcium and phosphorus intake provided by the ration in Table 5.22 was about 45 grams of each when no legume hay was fed. When alfalfa cubes were fed the average calcium intake was increased to 135 grams. So both diets supplied much more than the amounts needed by the maintenance horse and would easily replace the calcium and phosphorus lost in sweat.

All Thoroughbred and Standardbred trainers surveyed in the above study fed or injected a supplement such as wheat germ oil, vitamin-mineral mixtures, and solutions containing electro lytes and (perhaps) amino acids and vitamins. Electrolytes were given more often when the weather was warm than when it was cold.

We don't know the value of many supplements, and few controlled studies are available. Obviously, the addition of vitamins and minerals to deficient rations would improve performance, but megadoses to a balanced ration wouldn't do much. Selenium supplements, for example, would be required if the hay and grain were grown in selemum-deficient soils. Vitamin supplements would be of value if the hay was of poor quality, late cut, and had been stored for a long time.

"Blood builders" or hematinics are often used, but in a recent study hematinics did not change the resting red blood cell content in a healthy racehorse, and did not improve performance.

Iron supplements are often given to improve red blood cell levels, but deficiency is seldom a problem because most feeds contain a high level of iron. In a Cornell study horses were made anemic when their blood was drawn over a period of several weeks. All recovered at the same rate once the bleeding was stopped and they were fed a normal ration, or the ration plus chelated iron or the ration plus inorganic iron.

Dr. D. M. Keenan (1980) reported that packed cell volume (PCV), which indicates the number of cells per ml of blood, increases significantly in horses during a race. The average PCV in 86 horses was 40 percent before the race, 67 percent 10 minutes after the race, and 49 percent 1 hour after the race. The increase was considered to be due to the release of red blood cells from the spleen.

According to Dr. Keenan, the practice of "blood doping" to

increase the PVC doesn't improve performance because the increase in viscosity could have the opposite effect.

Dr. J. Sumner (1980) notes that vitamin and mineral supplements don't really affect athletic performance when athletes eat a balanced diet—and the same could be said of horses. But we live in a vitamin-crazed age, and many athletes continue to consume large quantities of supplements.

ENDURANCE HORSES

Of all the performance horses I suspect that the endurance horse is most likely to benefit from ration manipulation.

Studies with human athletes indicate that the use of electrolytes and glycogen loading are more likely to benefit long-distance performers than sprinters. Although glycogen loading might not be practical for endurance horses, high-energy diets may be beneficial. Studies with dogs by Kronfeld et al. (1977) demonstrated that high-fat diets increased the performance of sled dogs on extended runs.

We added 8 percent fat to the rations of horses going 50 miles at a rate of 9 miles per hour, but no clinical improvements were noted. However, the fat did help alleviate the decrease in blood glucose levels. It also decreased the ration needed, because fat contains more than twice as much energy as carbohydrates per pound and is highly digestible. In theory, a high-fat ration would be expected to cause a lower incidence of founder than a high-carbohydrate ration.

Studies by Slade (1979) indicated that horses fed high levels of protein may have decreased performance and sweat more than horses fed a control ration. They also require more water to excrete excess nitrogen in the urine. We found no differences in performance of horses fed 12 or 24 percent protein. Nevertheless, protein is not utilized as efficiently for energy as carbohydrate, so the horses fed the 24 percent protein required more feed. This raised costs because protein is more expensive than carbohydrate. We therefore concluded that there was no justification for feeding rations with more than 12 percent protein to endurance horses.

Minerals are lost in sweat. Electrolytes may be of particular concern for horses that sweat profusely, since synchronous diaphragmatic flutter (thumps) has been related to low calcium and potassium blood levels. Dr. Rose and co-workers (1980) calculated that the loss of electrolytes could easily account for a 30-percent decrease in plasma potassium and a 16-percent decrease in plasma chloride. Therefore, many endurance riders give their horses electrolytes during the event. However, Dr. Rose concluded that the composition of some commercial mixtures is too high in bicarbonate and too low in potassium.

The degree and amount of work and the degree of sweating will likely influence the response to electrolytes. We found no electrolyte problems in horses going 50 miles per day at 9 miles per hour, even when the temperature was 90°F. However, the terrain was not difficult and the horses were in excellent condition.

Drs. Lucke and Hall (1980) reported that in 1979 only 13 of 62 starters completed the Golden Horseshoe Ride (50 miles on day 1 and 25 miles on day 2). The maximum temperature on the first day of the ride was only 74°F, but it was the first warm day of the year and the horses had been trained in much colder weather. The plasma levels of potassium and calcium decreased significantly, and the animals became dehydrated and exhausted. An intravenous feeding of balanced electrolyte solutions was suggested to help improve recovery. However, there is no evidence to suggest that feeding high levels of electrolytes *prior* to the event will build up electrolyte stores in the horse's system.

Dehydration is one of the most common problems in endurance horses. In 1977 Hinton wrote, "The simplest and most obvious way to prevent dehydration is to allow horses to drink whenever they wish"—and this means every hour or so when they are working hard. Hinton suggested that horses on long rides will need an average of about a gallon of water per hour, from the start of the ride until the horses are fully rested. Horses that are not in condition may sweat more and require even more water. He also wrote that there is no danger of a horse drinking to excess at the end of a ride unless it is dehydrated and very thirsty. Then the water should be rationed to a gallon every 15 to 20 minutes.

All our studies suggest that proper conditioning and training will determine the success of an animal. No diet can overcome poor training!

DRAFT HORSES

The draft horse is making a major comeback in the United States. In recent years there has been a dramatic increase in its use for pleasure and work.

Morrison (*Feeds and Feeding*) suggested that a 1,200-pound draft horse doing hard work could be fed: 12 pounds of grass hay and 16 pounds of oats; or 12 pounds of alfalfa hay and $13^1/_2$ pounds of corn; or 12 pounds of grass hay, 13 pounds of corn, and 1 pound of a protein supplement such as soybean meal. These values would be increased by 50 percent for an 1,800-pound horse. The rations suggested by Morrison would provide about 34 Mcal of digestible energy daily for the 1,200-pound horse and 48 Mcal of digestible energy for the 1,800-pound horse.

The U.S.D.A. *Farmers Bulletin 1030*, published in 1919 during the peak years for draft horses, indicated that a 1,200-pound horse at severe hard work could be fed: 10 pounds of timothy hay, 6 pounds of clover hay, $14^1/_2$ pounds of oats, and 2 pounds of wheat bran; or 14 pounds of alfalfa hay, 14 pounds of corn, and 4 pounds of corn stover; or 10 pounds of alfalfa hay, 7 pounds of prairie hay, 12 pounds of barley, and 2 pounds of corn gluten meal. These rations supply slightly more energy than those suggested by Morrison.

The National Research Council's recommendation for a 1,200-pound animal of light horse breeding is 18 Mcal for maintenance and 20 Mcal per hour of strenuous effort (such as racing at full speed) for a total of 38 Mcal. A light horse working hard for one hour daily would require more energy than a draft horse of the same size working in the field for 6 to 8 hours per day (thus 38 vs. 32 Mcal). Of course, speed is very energy inefficient, and the rate of working will greatly influence energy requirements. Morrison concluded that draft horses perform work most efficiently when pulling at a speed of 2 to 2.5 miles per hour.

According to Morrison, a 1,200-pound horse at light work could be fed: 16 pounds of grass hay and 6 pounds of oats; or 16 pounds of grass hay, $4^{1}/_{4}$ pounds of corn, and $^{1}/_{2}$ pound of soybean meal; or 16 pounds of alfalfa and 4 pounds of corn. Such rations would supply about $22^{1}/_{2}$ Mcal daily.

Remember, the amount of feed actually given should be determined by the body condition of the animal rather than by these guidelines. Many factors influence the energy requirement of a horse.

Estimates of the daily feed intake for draft horses, based on early studies, are shown in Table 5.23. These values, however, are only guides. Many factors such as environment, quality of feed, individuality of the horse, type of terrain, difficulty of work, and skill of the driver all influence feed requirements. For example, a nervous rather than steady driver who hitches the horses improperly can greatly decrease the efficiency of the horses. The values are based on studies conducted in the 1930s and '40s. Work is assumed to be plowing and hauling when walking at a speed of 2 to 2.5 miles per hour. Increased speed decreases efficiency. The horse that is not working does not need any grain. In fact, a horse can become fat on good quality hay alone.

Numerous trials have demonstrated that draft horses can be fed corn and alfalfa rather than oats. This did not cause the draft horses to sweat more than when fed oats and timothy, and it was usually more economical.

The grain ration of a hardworking horse should be reduced by at least 50 percent on days of rest to decrease the incidence of "Monday morning" disease, or exertional myopathy.

Morrison (*Feeds and Feeding*) suggested that the daily amount of grain be divided equally into three feedings at morning, noon, and night. He also stated that one quarter of the hay allotment can be fed in the morning, one quarter at noon, and one half at night; alternatively, one third of the hay can be fed in the morning, none at noon, and two thirds at night. This gives the horse time to eat and digest the hay, and will prevent distention of the digestive tract when he is at work.

Protein requirement is not greatly increased by work. Mor-

Table 5.23. Estimates of Daily Feed Intake for Various Levels of Activity of Draft Horses[a]

Body Wt. (lb.)	No Work		Light[b]		Medium[c]		Heavy[d]	
	Hay (lb.)	Grain (lb.)	Hay (lb.)	Grain (lb.)	Hay (lb.)	Grain (lb.)	Hay (lb.)	Grain (lb.)
1,200	17–19	None	15–17	6–7	14–15	9–11	12–13	14–16
1,400	20–22	None	17–19	7–8	16–17	10–12	14–15	16–17
1,600	22–25	None	20–21	8–9	18–20	12–14	16–18	18–20
1,800	25–27	None	18–20	9–10	20–22	14–16	18–20	20–23

[a] Based on Morrison (*Feeds and Feedings*) and 1949 National Research Council Recommended Nutrient Allowances for Horses.
[b] 2 to 3 hours per day.
[c] 4 to 5 hours per day.
[d] 6 to 8 hours per day.

rison's suggested rations for heavy work contain 10 to 11 percent protein.

Early studies at the University of Illinois and Iowa State University showed that the draft horse will lose calcium in sweat but that either good quality grass hay or legume hay would supply the required amount.

Trace-mineralized salt should be fed free choice.

Morrison reported that pregnant draft mares can be worked. In fact, the exercise of regular work is beneficial; however, pulling too hard, backing heavy loads, and wading through deep snow or mud or other overexertion can be dangerous to both mare and fetus. He suggested that as foaling time nears, the work should be lightened and then discontinued 3 to 7 days before foaling. Even then, though, the mare should not be confined to a stall but given an opportunity to exercise. Moreover, the mare should always be handled gently. Draft mares were often returned to light work within 10 days after foaling if the foaling was not difficult.

Donkeys and Mules

The National Research Council does not provide any information about the nutrient requirements of donkeys and mules, but mules may require less enregy than horses to do the same amount of work.

Two teams of horses (Charley and George; Maude and Fanny) were compared to two teams of mules (Peter and Kit; Logan and Joe) in trials conducted by Professor C. M. Conner at the Florida Experiment Station in 1904. The horses weighed an average of 1,050 pounds and the mules weighed an average of 975 pounds. The animals were used for hauling and plowing for three months. The average daily intake of the horses was 14 pounds of hay and $9^1/_2$ pounds of grain compared to 9 pounds of hay and 10 pounds of grain for the mules. In another trial the horses ate 5 pounds of corn, 13 pounds of hay, and $13^1/_2$ pounds of sweet potatoes compared to 5 pounds of corn, 9 pounds of hay, and 14 pounds of sweet potatoes for the mules. The average daily digestible energy of the horses for both trials was 29 Mcal compared to 26 Mcal for the mules—a difference of about 10 percent.

A similar study at the University of Illinois also indicated that mules would require about 10 percent less energy than horses. The animals were worked an average of 8 hours per day.

Donkeys are also reputed to be more efficient than horses. There is a traditional saying that donkeys can exist on thistles and straw. True, they can exist in areas not suitable for horses, but their advantage may be in conservation and grazing behavior rather than increased efficiency of energy utilization. Further studies are needed. In preliminary studies we found no large differences between horses and donkeys in the ability to digest traditional horse feeds. However donkeys appear to tolerate water deprivation better than horses.

I do not know of any studies on nutrition of hinnies (the offspring of the mating of a stallion and a jennet). Presumably their requirements would be similar to those of mules when expressed on a weight basis. Hinnies are usually slightly smaller than mules.

References

Cahill, C. Presented at N.Y. Horse Breeders Conference, Albany, 1981.

Conner, C.M. Feeding horses and mules. *Florida Ag. Exp. Sta. Bull.* 72, 1904.

Dadd, G.H. *Amer. Vet. J.* 1:434, 1855.

Francis-Smith, K. and D.G.M. Wood-Gush. Coprophagia as seen in Thoroughbred foals. *Eq. Vet. J.* 9:155, 1977.

Haycock, W. *The Gentleman's Stable Manual.* London: Routledge, Warne and Routledge, 1861.

Herbert, H.W. *Hints to Horse-Keepers.* New York: Orange Judd Co., 1881.

Hinton, M. Long distance horse riding and the problem of dehydration and rhabdomyolosis. *Vet. Ann.* 17:136, 1977.

Houpt, K. Personal communication, 1981.

Hudson, R.S. Feeding draft colts. *Proc. 26th Meeting Amer. Soc. Anim. Prod.* p. 104, 1933.

Jeffcott, L.B. Some practical aspects of the transfer of passive immunity to newborn foals. *Eq. Vet. J.* 6:109, 1974.

Keenan, D.M. Changes in packed cell volume of horses during races. *Aust. Vet. Pract.* 10:184, 1980.

Kronfeld, D.S. Feeding on horse breeding farms. *Proc. Amer. Assoc. Eq. Pract. St. Louis* p. 461, 1978.

Kronfeld, E. et al. Responses in racing sled dogs. *Amer. J. Clin. Nutr.* 30:419, 1977.

Lathrop, A.G. and G. Bohstedt. The use of oat feed as the entire ration for horses at light to medium work. *Proc. 25th Meeting Amer. Soc. Anim. Prod.* p. 102, 1932.

Lucke, J.N. and G.N. Hall. Further studies on the metabolic effects of long distance riding: Golden Horseshoe Ride 1979. *Equine Vet. J.* 12:189, 1980.

Mason, T.A. A high incidence of congenital angular limb deformities in a group of foals. *Vet. Rec.* 109:93, 1981.

Miller, D. Feeding, in *Care and Training of the Trotter and Pacer.* J.C. Harrison, ed., Columbus, Ohio: USTA, 1968.

Roberts, I.P. *The Horse.* New York: Macmillan Co., 1913.

Rose, R.J. et al. Plasma and sweat electrolytes concentrations in the horse during long distance exercise. *Eq. Vet. J.* 12:19, 1980.

Simms, J.A. and Williams, J.O. Hay requirements of city work horses. *Conn. Agr. Coll. Bull.* 173, 1931.

Slade, L. Protein needs of equine athletes. *Utah Sci.* 40:10, 1979.

Stewart, J. *The Stable Book.* Edinburgh: Blackwood and Sons, 1841.

Stromberg, B. A review of the salient features of osteochondrosis in the horse. *Equine Vet. J.* 11:211, 1979.

Sumner, J. Nutrition and athletic performances. *Food Nutr.* 37:125, 1980.

Tyler, C. The development of feeding standards for livestock. *Ag. Hist. Rev.* 4:97, 1957.

Tyznik, W. Nutrition, in *Care and Training of the Trotter and Pacer.* J.C. Morrison, ed., Columbus, Ohio: USTA, 1968.

Williams, J. Presented at N.Y. Horse Breeders Conference, Albany, 1981.

Winter, L.D. A Survey of Feeding Practices at Two Thoroughbred Racetracks. M.S. thesis, Cornell University, Ithaca, N.Y., 1980.

Chapter Six

Metabolic Problems and Other Concerns

I'd like to deal with some metabolic problems and other matters of concern to horsemen that have not yet been discussed. Topics will include founder, exertional myopathy, hyperlipemia, eclampsia, epistaxis, heaves, hoof growth, sweet feed bumps, hair coat, and termperament and behavior.

Founder

Founder (laminitis) is a metabolic derangement. It causes severe pain, inflammation, and separation of the sensitive from the insensitive lamina of the hoof. The problem has been studied for centuries but it is still poorly understood and hotly debated.

Founder can be classified as acute or chronic. Full-blown symptoms appear quite suddenly in horses with acute founder. The horses are reluctant to move forward; when forced, they walk heel to toe as though "walking on eggs." The foot will feel warm to hot, and the sole ahead of the apex of the frog will be very tender. In severe cases the coffin bone may rotate and protrude through the sole. The digital pulse may be pounding. Temperature may increase to 106°F and respiration to 50 to 70 breaths per minute.

Obel (1948) classified founder into four grades:

Grade 1. In the standing position, the horse lifts its feet incessantly, often at intervals of a few seconds ("paddling"). At a walking pace it doesn't show any lameness, but the trotting gait is short and stilted. This continues when the horse is recovering.

Grade 2. The horse moves quite willingly at a walking pace, but the gait is stilted. A forefoot may be lifted without difficulty.

Grade 3. The horse moves most reluctantly, vigorously resisting any attempt to have a forefoot lifted.

Grade 4. At this stage the horse does not move without being forced.

After the initial inflammation subsides (which may require one to several days), the horse may have chronic founder. The coffin bone rotates, causing widening of the white line and abnormal growth rings on the hoof. The distance between rings is usually greater at the heel or quarter than at the toe, which will curl upward like a babouche (Turkish slipper) if not trimmed regularly.

Some of the causes of founder have been known for years. In 1880 James Law, the first professor of veterinary medicine at Cornell University, wrote that founder could be a consequence of giving large quantities of cold water to a hot horse, of hard road work, pneumonia, metritis (inflammation of the uterus) and retained placenta, indigestion, or a grain overload.

However, the treatments for founder in 1880 were different from treatments of today. Dr. Law suggested that horses be given sedatives such as lobelia (wild tobacco), aconite (wolf's bane), or tobacco; he also advocated wrapping the hoof with a warm poultice. If the feet were not too tender, he suggested that walking the horse on a soft, newly plowed field would improve circulation. Law wrote, "In severe cases the coronet may be scarified with a sharp lancet and the foot placed in a bucket of warm water to favor bleeding."

Fat ponies and horses that are easy keepers are much more likely to develop founder than animals in trim condition. Of the 5,911 Thoroughbreds, Quarter Horses, Arabians, and Standardbreds admitted to the Cornell Clinic over a 6-year period, only 3.0 percent had founder; 6.0 percent of the 251 Morgans had founder, perhaps because Morgans are relatively easy keepers. The incidence in the 345 ponies was 5.2 percent.

Dr. H. Garner (1980) concluded that female and male horses show an equal risk for founder, but that 4 to 7 years of age and 7 to 10 years were the highest risk groups for females and males respectively. He also said that the ratio of total body mass to hoof is a significant factor; thus the heavily muscled Quarter Horse with a small hoof is a potential candidate for severe founder.

Animals—particularly ponies—suddenly given free access to lush pasture are prone to founder. Therefore, introduce your animals gradually to pasture.

When the horse suddenly obtains a large amount of grain, the relative amounts of certain bacteria in the large intestine are likely to change, producing increased amounts of lactic acid. This may somehow trigger the metabolic change leading to founder. Perhaps the acid dissolves the wall of other bacteria and releases an endotoxin. The signs of grain founder may appear within a few hours after ingestion, but frequently do not show until 12 to 18 hours later.

The message is clear. Acute or chronic overfeeding of grain causes founder. When increasing grain intake, do it gradually to give bacteria an opportunity to adapt. Make certain that feed bins are covered so that if the horse gets out of his stall he cannot gorge himself.

Chronic overfeeding of grain may be a particular problem in show horses. Garner reported that the highest risk for laminitis in Quarter Horses was in the late summer and fall—or the peak of the show season—when horses have been fed high levels of grain to get them in show condition.

Garner also pointed out that founder may result from stress combined with the genetic and metabolic "set-up." He wrote, "The single and combined stresses of changes in either external or internal environments should be avoided or lessened when possible. A 4- to 10-year old 1,400-pound Quarter Horse, standing on double-0 shoes, maintained on a diet with a high ratio of grain to hay, and frequently trailered in extremely hot or cold weather, is at an unnecessarily increased risk for developing laminitis."

Garner suggested that it was not necessary to feed high amounts of grain to keep horses in show condition if a high quality hay was fed.

Horses that have foundered are predisposed to founder again.

Another cause of founder that has received considerable attention recently is the bedding of horses on freshly planed black walnut shavings. Dr. True and co-workers (1978) reported that on 7 farms, 147 of 232 horses developed founder within 24 hours after being bedded on black walnut shavings. The founder in many of the horses was so severe that the usefulness of the animals was greatly impaired. Treatment for founder on some farms was sometimes delayed because other disease conditions were suspected. Founder usually hits the individual and is seldom a herd problem.

The agent in black walnut shavings responsible for inducing founder has not been identified, but a suspect compound is a naphthaquinone called juglone. Earlier reports have shown juglone to be toxic to plants like tomatoes and alfalfa. It is also toxic to earthworms. Studies are now in progress to determine if juglone is toxic to horses.

It is possible that the toxic fraction in walnut shavings is destroyed during storage. In the seven farms studied, the shavings were freshly planed.

Exertional Myopathy

Exertional myopathy is a clinical problem with many names. It has been called azoturia, Monday morning disease, paralytic myoglobinuria, exertional rhabdomyolysis, tying-up syndrome, and myositis.

The signs may include pain and stiffness of the back and thighs. The animal may refuse to move or even assume a dog-sitting position. Sweating may be apparent, particularly in the areas of the affected muscles. The urine may be coffee-colored due to the presence of myoglobin released from the damaged muscle cells. The blood levels of several enzymes such as serum glutamic oxalacetic transaminase (SGOT), lactate dehydrogenase (LDH), and creatine phosphokinase (CPK) are also increased because of their release from damaged tissue.

Fillies are more susceptible to exertional myopathy than colts. Heavily muscled animals such as Quarter Horses and Draft Horses may have a greater incidence than horses with lighter muscles. Nervous animals and animals exercised in damp and cold conditions also tend to develop exertional myopathy.

The condition often occurs in hardworking horses fed a high-grain diet but then rested for one or more days with no reduction in feed. Therefore, grain intake should be reduced by 50 percent or more on days of rest. The mechanisms by which the high grain intake triggers the condition is not known, but a buildup of glycogen is often involved.

Selenium and vitamin E injections have been widely used to prevent exertional myopathy, although they are not always effective. Dr. Karl White (1982) suggested that in mild cases the horse should be kept warm (by blankets if necessary) and walked slowly. But if the pain becomes severe, walking should be stopped immediately; horses with severe myopathy should not be moved.

The veterinarian may use phenylbutazone to relieve pain. Tranquilizers may also be useful. Fluid therapy may be necessary.

The animal should be kept warm in a quiet environment and given only hay and water. He should be rested for several days after the attack.

Dr. White concluded that the plane of nutrition must correlate on a daily basis with the level of exercise. Consistency is important. He suggested that weekend riders often have difficulty giving their horses a consistent feed and exercise level. If exertional myopathy is a problem for such an owner he might consider acquiring a horse less sensitive to variations.

Incidentally, the incidence of exertional myopathy seems to be increasing in urban cowboys, according to a report in the *New England Journal of Medicine*. Inexperienced and out-of-condition riders that try to conquer mechanized bulls may develop myopathy, have severe thigh cramps, and produce dark red urine. However, they usually recover without complications after a few days of rest. (Of course, orthopedic injuries are much more common than myopathy in "riders" of this sort.)

Hyperlipemia

Hyperlipemia means there is a greater than usual level of fat (particularly triglyceride) in the blood. The serum or plasma of affected horses will appear cloudy or creamy.

The problem arises when animals—particularly fat ponies—go off feed. The body fat is mobilized to provide energy. But in some cases the balance between mobilization and utilization is not maintained and excessive fat accumulates in the blood. The normal range of total lipids is 100 to 500 mg per 100 ml of blood, but the average of total lipids in 7 animals with hyperlipemia was 1,760 mg (Naylor et al., 1980).

The problem occurs most frequently in overweight ponies either under the stress of late gestation or at peak lactation that have sudden food deprivation, or in animals that become anorexic (go off feed) secondary to an illness. But the condition can occur in any class of horse or pony.

The reason for the anorexia should be treated promptly. In those animals that become very lipemic, the prognosis is guarded at best. Glucose and heparin or glucose and insulin treatments sometimes help decrease the fat level in the blood, but it is critical that the animal resume eating.

Naylor et al. suggested that offering small quantities of a selection of freshly prepared feeds, particularly grass, may entice the animals to eat. If this fails, they suggest that the animal be force-fed by a stomach or esophagostomy tube. In the latter case the tube is inserted through a surgical opening in the esophagus and kept in place so that small amounts of feed can be given frequently without the dangers of frequent use of the stomach tube.

In our studies hyperlipemic ponies were more likely to eat poor quality grass than they were to eat alfalfa hay or grain, but the idea of providing a selection of feeds seems to be a good one.

Eclampsia

Eclampsia resembles milk fever in dairy cows. It is characterized by hypocalcemia (a low level of blood calcium) and is induced by the loss of calcium in milk. The signs include muscular tensions and tremors, anxiety, sweating, dilation of the nostrils, salivation, grinding of the teeth, and thumps (synchronous diaphragmatic flutter, similar to the condition already discussed in endurance horses). The calcium level of affected horses is usually 4 to 6 mg per 100 ml of plasma; the normal range is 11 to 13.5 mg. The blood level of magnesium may also be decreased, although in some cases it is increased. Phosphorus blood levels may be normal, increased, or decreased.

Eclampsia is generally considered to be less of a problem now than when draft horses were more prevalent; however, cases are reported. The disease most commonly occurs around the tenth day after foaling but may occur in mares in mid-gestation and mid-lactation. Added stress such as rounding up and penning the animals, hard physical work, and prolonged transport may precipitate the condition.

Dr. J. Arngjerg (1980) reported that treatment with an intravenous injection of a 500-ml solution containing 8.42 grams of calcium, 1.88 grams of magnesium, and 4.8 grams of phosphate has been effective. In some cases, he suggested that two injections may be required. The mares usually respond dramatically, but a few may take several days to recover.

Injections of excessive amounts of calcium can cause heart failure.

Epistaxis

Epistaxis is a fancy word for bleeding from the nose. It was formed from the Greek *epi*, meaning "on," and *staze*, meaning "to fall in drops." Although many people think that a "bleeder" horse has broken a blood vessel in the nasal cavity (as happens in a person), the blood from a horse with epistaxis usually comes from hemorrhage in the lungs. When pulmonary hemorrhage occurs in humans, the blood is coughed up and spat out of the mouth. However, the horse has a long soft palate, so the blood is discharged from the nostrils.

Dr. W. R. Cook (1974) reported that bleeders were usually racehorses, but hunters and event horses, polo ponies and unworked horses could also be bleeders. In his study, the age of affected horses ranged from 1 to 13 years of age; the majority, however, were 6 to 9 years old. Horses that were overweight or not in condition were more likely to develop the condition than horses in good condition.

Cook recommended that horses with a history of epistaxis be housed in a well-ventilated stable. Exposure to dust should be reduced as much as possible. Wood chips are preferred to straw for bedding. Feed the horse pellets or cubes rather than hay to reduce dust, or immerse the hay in water prior to feeding.

Racehorses in Australia that bled when stabled did not bleed when trained from pasture, suggesting the influence of dust on epistaxis. Cook also suggested that bleeders have their hay ration reduced the night before a race, and that they not be fed any hay on a race day, in order to reduce pressure exerted on the diaphragm by the abdominal contents. Likewise, the thorax should not be obstructed by excessive tightening of the girth.

Dr. J. R. Pascoe and co-workers (1981) recently examined 235 horses with a flexible fiberoptic endoscope at the Del Mar Racetrack. Only 2 (0.8%) had blood flow from the nostrils, but 103 (43.8%) had various degrees of hemorrhage in the tracheal lumen; the blood appeared to come from the lungs. They suggested that the term "exercise-induced pulmonary hemorrhage" was more accurate than "epistaxis," "bleeding," and "sporadic idiopathic epistaxis." Although the drug furosemide is widely used to treat the condition, they also concluded that it may not be very effective in the prevention of pulmonary hemorrhage.

Vitamin C has also been used to treat this condition. However, the vitamin C levels in the blood of horses with epistaxis were not different from those of horses without the problem. A deficiency of vitamin K would greatly aggravate the problem, but the horse appears to obtain ample vitamin K from bacterial synthesis in the digestive tract.

Heaves

Heaves (pulmonary emphysema or broken wind) is the accumulation of air in the lungs. That is, the air cannot be expired in a normal manner. The disease is characterized by chronic cough, difficult breathing (shortness of wind), lack of stamina, and—perhaps—intermittent nasal discharge.

Horses that have had heaves for an extended period will have hypertrophy (increased size) of the external abdominal oblique muscles (heave line), because these muscles are used in an attempt to force the air from the lungs.

The specific causes of heaves are not known, but it has been suggested that allergies might be involved. Attacks are more frequent during hot, humid weather and in dusty environments. Horses out on pasture almost never have heaves. (Other methods of reducing dust were discussed in the section on epistaxis). Complete pelleted feeds containing fiber sources other than hay, such as beet pulp, are often fed to horses with heaves because it is thought that some horses may be allergic to proteins in hay.

Horses with heaves are often used as a model for studies on emphysema, because some cases of emphysema in horses are similar to the problem in humans.

Hoof Growth and Nutrition

What is the relationship of nutrition to hoof growth? Can hoof quality and hoof growth be improved by proper feeding? Of course, fast growth is not necessarily associated with high-quality hooves, but it is useful when growing out a hoof with a problem such as a quarter crack. Fast growth may also be helpful for endurance horses or racehorses that must have their shoes replaced frequently. Horses with thin walls or brittle hooves are likely to have hoof diseases.

The first step is to select horses that have the genetic potential to have good hooves. I could not find any studies on this, but I suspect that the heritability of hoof growth and quality is high. Studies with cattle have demonstrated that selection for rate of hoof growth, heel depth, or hoof composition is possible through heritability.

Hoof growth is also influenced by the age of the animal. We found that foals under nine months of age had a growth rate of 0.5 mm per day, whereas 0.2 to 0.3 mm was found in mature animals under similar conditions. Shannon and Butler (1979) found that horses two and three years of age had a faster rate of hoof growth than horses seven years or older.

In most situations hind hooves grow at a faster rate than front hooves. However, several factors such as use, age, and conformation may influence the difference. In some cases, there may be no difference between hind and front, and sometimes the front may even grow slightly faster than the hind.

Hooves grow faster at different times of the year—influenced perhaps by temperature and weather. The hooves of polo ponies at Cornell grew faster in September and October than during January and February. Shannon and Butler reported that the average rate of growth for horses in western Texas was greater in October (.27 mm/day) than in December (.19 mm/day). They suggested that the higher temperature and greater moisture in October may have been influential.

It's possible that increasing the blood supply to the foot will increase rate of hoof growth. Thus irritants, blisters, or other circulation-stimulating agents are sometimes applied to the coronet. It should be noted, though, that some practitioners believe the irritants may result in hoof tissue of poor quality. Daily massage of the coronet with the hand or with a toothbrush has also been reported to increase circulation and rate of hoof growth.

Reduced energy intake can result in decreased rate of hoof growth in growing animals. A complete pelleted ration was fed limited or free choice to two groups of Shetland pony foals for 117 days (Butler and Hintz, 1977). The limited group gained about 0.2

pound per day. The free choice group gained 1 pound per day. Hoof growth in the limited group was .25 mm per day, compared to .38 mm per day in the free-choice group. Of course, it might be possible that any factor decreasing body weight gain will decrease rate of hoof growth. Several studies have shown a high correlation between rate of hoof growth and rate of weight gain.

In other Cornell studies protein deficiency was shown to decrease hoof growth. The hoof growth of weanlings fed 9 percent protein was only two thirds of weanlings fed 14.5 percent protein.

Several studies have been conducted to determine whether additional protein or amino acids accelerate hoof growth. We found no benefit from the addition of gelatin to commercial, complete pelleted feed. Hoof growth was .33 mm per day for those weanling ponies fed the pellets and .31 mm per day for those ponies fed the pelleted diet plus 90 grams of gelatin per 100 kg of body weight.

Increasing the protein content of rations for weanlings from 14.5 percent to 22 percent did not increase rate of growth either. Don't expect the addition of amino acids to balanced rations to improve the rate of growth of the hoof. Schott et al. (1981) studied the effect of supplemental methionine on polo ponies. The ponies (actually horses with primarily Thoroughbred breeding) were divided into three groups. Group one was fed the control diet of 13 pounds of a commercial pelleted feed and 6 pounds of timothy-alfalfa hay (mostly timothy). Group two was fed the control diet plus 3 grams of DL-methionine per day. Group three received 6 grams of DL-methionine per day. The methionine was mixed with molasses and poured over the pellets. The trial lasted 210 days, and no significant differences in growth rate were found. The average rate was approximately .20 mm for all three groups.

The addition of 0.3% lysine or 0.3% lysine and 0.1% methionine to a ration containing 14% protein did not improve the rate of hoof growth of weanling horses. The average daily rates were .62, .61, and .60 mm for the control, the control plus lysine, and the control plus lysine and methionine respectively (Meakim, 1979).

Richardson and Ott (1977) reported that the rates of hoof growth of yearlings fed either a control ration or a control plus 0.15% lysine or plus .30 lysine were .40 and .41 mm per day respectively.

Horses suffering from vitamin A deficiency may have marked scaling of the periople. Biotin deficiency has been reported to cause heel cracks, erosion of the heel, cracks at the junction of heel and toe, and cracks in the toe itself in swine. It would probably cause hoof problems in horses, but biotin deficiency in horses appears to be highly unlikely.

Zinc is essential for normal epidermis. In a recent symposium it was suggested that zinc deficiency could be a cause of foot problems in cattle. Although zinc deficiency has not been reported under field conditions in horses, such a possibility cannot be ignored. It was previously thought that zinc deficiency would not be a problem in ruminants.

Selenium toxicity in horses can result in hair loss from the mane and tail and degeneration of hoof quality, eventually leading to loss of the hoof.

Sweet Feed Bumps

Many horses develop small bumps or swellings of about .5 mm in diameter in the skin, often in the saddle and girth area. The bumps are noncontagious.

The cause is not known. Many people think that the condition is an allergic response to feeds—hence the name "sweet feed bumps." Others feel that protein is the problem, so they call them "protein bumps." I do not know of any evidence demonstrating the role of dietary ingredients in the development of the bumps. However, they are usually seen on horses that are fed a high plane of nutrition.

Many other causes, such as allergy to biting insects, photosensitization, and onchocerciosis (a filarial nematode) have been suggested. The bumps are most common in the warmer months, so they are sometimes called "heat bumps." The seasonal occurrences suggest that insects may be more of a factor than feed.

Dr. J. E. Lowe (1982) reported that bumps are discouraging to both the veterinarian and horse owner; they seldom get better and seldom get worse, in spite of treatment. Steroids may suppress the bumps, but they reappear when treatment stops. The horse owner may have to learn to tolerate the bumps until cold weather arrives.

Hair Coat

The condition of the hair coat often reflects the health status of the animal. Heavily parasitized animals or animals with hormone insufficiency—especially of the thyroid—often have a rough hair coat. Protein deficiency can result in a coat of poor quality; on the other hand, feeding levels of protein above the requirements do not produce a super coat. Animals fed copper-deficient diets have faded coat colors. Deficiencies of B-vitamins cause dermatitis in most species—but again, high levels of supplements do not appear to be of value.

Boiled flax seed has long been used to promote a bloom on an animal, possibly because of the mucin and oil contained in the seed. Unsaturated fats and oils such as corn oil are also frequently used. A dose of 2 or 3 tablespoons of corn oil daily seems to be a reasonable amount; it can take 30 to 60 days before a response is noticed.

The shedding of hair can be affected by increased daylight. For example, mares that are put under lights to induce ovulation often shed earlier than other mares. The most usual procedure is to subject the animal to a total of 16 hours of light starting in November or December.

Sometimes old advice is best. Recommendations for the improvement of hair coat were written in 1856 in *The American Veterinary Journal* (vol. 1, p. 285) by an author called Horse Latinus:

Brushus of curricomus	ad-libitum
Elbow greesus	quantum-suff
Blanketisus	first-ratus
Fodderus	Never-say-diet-us-but-meal-us-et-oatus
Exercisus	non-compromisus
Results: Coatus shinitus	

Temperament and Behavior

A horse fed an adequate level of energy will be livelier than a horse fed a low level of energy. "A horse feeling his oats" has energy; an underfed horse will be depressed and easier to handle. Some horse owners request a feed that will make their horse look good but will not make him "high." Nevertheless, energy is energy. I don't know of a feed that provides energy for body growth but not for activity.

Of course, a deficiency of any nutrient can cause depression. Are there any feeds or nutrients that can cause changes in temperament or behavior if given in large amounts? Early studies indicated that injections of 1,000 mg of thiamin had a tranquilizing effect on racehorses, whereas oral doses of up to 2,000 mg produced no observable effects. (The National Research Council's suggested adequate level is 40 mg.) However, in a subsequent study by Irvine and Prentice (1962) injections of up to 2,000 mg of thiamin had no effect on the pulse rate and no tranquilizing effect. They concluded that the earlier studies were not properly controlled. Stewart (1972) also reported that an injection of 2,500 mg of thiamin had no significant effect on the heart rate, respiration rate, or speed of the Thoroughbred, but on some occasions they were less excitable.

Protein deprivation has been shown to decrease learning ability in several species. By the same token, high levels do not create a super brain. Kratzer et al. (1977) studied the maze learning ability in Quarter Horses fed diets containing 10, 13, 16, or 19 percent protein. There were no differences in the ability to escape from an open maze or from the right side, but the horses fed 10 percent protein discovered the left side escape sooner. The researchers felt it was unwise to draw any hasty conclusions until further studies were done.

Anemia

Anemia, or lack of hemoglobin, can be caused by many factors. Deficiencies of any of several nutrients such as iron, copper, vitamin

B12, folic acid, vitamin B6, vitamin E, and protein can be causes. But there is no evidence to suggest that supplementing a balanced diet with any of these nutrients will increase the blood cell count.

Anemia can be caused by blood loss due to injury, internal parasites, or external parasites. When I read Louis L'Amour's account in *Lonely on the Mountain* of mosquitoes killing horses by draining them of blood, I thought the author was exaggerating. However, scientific reports now show that large numbers of mosquitos can be fatal. Drs. Bruce Abbitt and Lauren Abbitt recently reported such a problem in a herd of 1,200 cattle on low-lying pastures bordering salt marshes along the San Bernard River in Brazona County, Texas. Swarms of feeding mosquitoes were seen to be bothering the cattle. Within three days several animals had died—some within 12 hours after the mosquitoes were first observed. Laboratory tests indicated that the anemia was severe enough to be the sole cause of death.

The Abbitts calculated that it would take about 3.8 million mosquito bites to remove 50 percent of the blood from a cow and that such an attack rate was possible. Fortunately, such severe attacks of mosquitoes are rare.

Other causes of anemia include immune destruction, such as in neonatal isoerythrolysis or in destruction of red blood cells by intracellular viral or rickettsial agents. The most common example is equine infectious anemia. Ingestion of toxic materials such as red maple leaves and wild onions can also cause anemia.

References

Abbitt, B. and L. Abbitt. Fatal exsanguination of cattle attributed to an attack of salt marsh mosquitoes. *JAVMA* 197:1397–1399, 1981.

Arnbjerg, J. Hypocalcemia in the horse. *Nordisk. Vet. Med.* 32:207–211, 1980.

Butler, K.D. and H.F. Hintz. Effect of level of feed intake and gelatin supplementation on growth and quality of hoofs and ponies. *J. Anim. Sci.* 44:257–261, 1977.

Cook, W.R. Epistaxis in the racehorse. *Eq. Vet. J.* 6:45–58, 1974.

Garner, H.E. Update on equine laminitis. *Vet. Clinics North Amer.* 2:25–32, 1980.

Irvine, C.H.G. and N.G. Prentice. The effect of large doses of thiamine. *New Zealand Vet. J.* 10:86–88, 1962.

Kratzer, D.D. et al. Maze learning in Quarter Horses. *J. Anim. Sci.* 45:896–902, 1977.

Law, J. *The Farmer's Veterinary Adviser.* Ithaca, N.Y.: Finch and Apgar, 1880.

Lowe, J.E. 1982. Heat Bumps, in *Current Veterinary Therapy for Horses.* Philadelphia: Saunders, 1982.

Meakim, D.W. M.S. thesis, University of Florida, Gainesville, 1979.

Naylor, J.F., D.S. Kronfeld, and H. Acland. Hyperlipemia in horses: Effects of undernutrition and disease. *Amer. J. Vet. Res.* 41:899–905, 1980.

Obel, N. Studies on the histopathology of acute laminitis. Almqvist and Wiksells Boktryckeir ab Uppsala, 1948.

Pascoe, V.R. et al. Exercise-induced pulmonary hemorrhage in racing Thoroughbreds: A preliminary study. *Amer. J. Vet. Res.* 42:703–707, 1981.

Richardson, G.L. and E.A. Ott. Influence of protein source and lysine intake on growth and composition of hoof yearling foals. *J. Anim. Sci.* 45 (Supplement 1):105, 1977.

Richardson, G.L. M.S. thesis, University of Florida, Gainesville, 1978.

Schott, H.C., H.F. Hintz, and S. Kraus. The effects of DL-methionine supplementation on hoof growth and hoof quality in the horse. Cornell University. Unpublished data, 1981.

Shannon, R.O. and K.D. Butler. Influence of age, season and hoof location on equine hoof growth. *Am. Farrier J.* 5:44–45, 1979.

Stewart, G.A. Drugs, performance and responses to exercise in the racehorse. *Aust. Vet. J.* 48:544–547, 1972.

True, R.G., J.E. Lowe, J. Heissen, and W. Bradley. Black walnut shavings as a cause of acute laminitis. *Proc. Amer. Assoc. Eq. Pract.* 511–516, 1978.

White, K.K. Exertional Myopathy, in *Current Veterinary Therapy for Horses.* Philadelphia: Saunders, 1982.

General Bibliography

Bradley, M. *Horses*. New York: McGraw-Hill, 1981.

Cunha, T. J. *Horse Feeding and Nutrition*. New York: Academic Press, 1980.

Ensminger, M. E. *Horses and Horsemanship*. Danville, Ill.: Interstate, 1969.

Evans, J. W., Borton, T., Hintz, H. F., and D. Van Vleck. *The Horse*. San Francisco: Freeman, 1977.

Gay, C. W. *Productive Horse Husbandry*. Philadelphia: Lippincott, 1932.

Kays, J. M. *The Horse*. Third edition, revised by John M. Kays. New York: Arco, 1982.

Lewis, L. *Feeding and Care of the Horse*. Philadelphia: Lea and Febiger, 1982.

Lowe, H. and H. Meyer. *Pferdezucht and Pferdefutterung*. Stuttgart, Germany: Verlage Eugen Ulmer, 1977.

Maynard, L. A., Loosli, J. K., Hintz, H. F., and R. G. Warner. *Animal Nutrition*. New York: McGraw-Hill, 1979.

Morrison, F. B. *Feeds and Feeding*. Ithaca, N.Y.: Morrison Publ., 1957.

National Academy of Sciences. Nutritional Data on United States and Canadian Feeds. Washington, D.C., 1971.

National Research Council. Nutrient Requirements of Horses. No. 6. National Academy of Sciences, Washington, D.C., 1978.

Robertson, E. I. In *Feed Industry Red Book*. Edina, Minn., 1981.

Willoughby, D. P. *Growth and Nutrition in the Horse*. Cranbury, N.J.: A. S. Barnes, 1975.

Index

Alcohol, 136
Alfalfa
  feeding value, 102, 117
  history, 3
  meal, 132, 139
Amino acids
  absorption, 15
  essential, 37
  in large intestine, 17
  requirement, 35
Ammonia, 17
Anemia, 52, 62, 72, 220
Antibiotics, 146, 147, 193
Arsenic, 65
Ascorbic acid, *See* Vitamin C
Axoturia, 210

Bahiagrass, 117
Bakery products, 136
Barley
  feeding value, 127
  history, 1
Beet pulp, 133
Behavior, 220
Bermuda grass, coastal, 110
Bile, 16
Biotin, 77
Birdsfoot trefoil, 106, 117
Bleeding, nose, 213
Blister beetles, 105
Bluegrass, 118
Body weight, 184, 188
Bone, 41, 186
Bone meal, 46

Bracken fern, 72
Brewers yeast, 72
Bumps, heat, 218
Bumps, sweet feed, 218

Calcium, 41
  deficiency, 42
  eclampsia, 213
  evaluation, 44
  oxalate, 44
  requirements, 41
  self-feeding, 47
  sources, 46, 105
  supplements, 46
Calcium to phosphorus ratio, 46, 104
Calorie, 33
Canary grass, 118
Canola meal, 178
Carbohydrate, 31
Carrots, 131
Cecum, 18
Cellulose, 31
Cell walls, 31
Chloride, 50
Choke, 14
Choline, 74
Chromium, 40
Citrus pulp, 133
Clover
  alsike, 117
  crimson, 107
  ladino, 117
  red, 106, 117

Clover (*continued*)
  sweet, 70, 107
  white, 117
Cobalt, 63, 72
Cocoa shells, 133
Cold, effect on energy of, 35
Colic, 21, 24
Colitis X, 26
Colon, 17
Colostrum, 178
Commercial feeds, 140
Condition scores, 35
Copper, 62
Corn
  feeding value, 122
  history, 1
Cottonseed meal, 137
Creep rations, 183
Cubes, 145

Dallisgrass, 118
Diarrhea, 27
Dicumarol, 70, 107
Digestibility of feeds, 18
  factors affecting, 18
Digestive system, 12
Distillers grains, 134, 139
Donkeys
  feeding, 204
  watering, 79
Draft horses, feeding, 201
Dried bakery products, 136
Dust control, 143, 213, 215

Eclampsia, 213
Electrolytes, 103, 200
Emphysema, 215
Energy, 31
  digestible, 33
  metabolizable, 33
  net, 33
  requirements, 159
Endurance horses, 199
Enterolith, 23
Epiphysitis, 186

Epistaxis, 213
Ergot in rye, 128
Esophagus, 14
Ether extract, 31
Exercise induced pulmonary
  hemorrhage (EIPH), 214

Fats and oils
  digestion, 16, 131
  endurance horses, 199
  energy source, 32
  feed content, 130
Fatty acids
  essential, 32
  volatile, 17
Feed intakes
  draft horses, 203
  growing horses, 193
  mares, 168
  racehorses, 197
Feeding programs, 154
Fescue toxicity, 119
Fences, rubber, 23
Fiber
  acid detergent, 31
  crude, 31
  neutral detergent, 31
Fishmeal, 140
Fluorine, 63
Foals
  newborn, 178
  orphans, 180
  suckling, 181
Folic acid, 73
Forages, 84
  composition, 86
  sampling, 94
Founder, 206

Gallbladder, 16
Gelatin, 217
Goiter, 58
Gossypol, 138
Grain sorghum, 127
Grains, 121
  standards, 125

Grass tetany, 48
Grasses, 85, 108, 117
Growth rate, 184, 188

Hair coat, 219
  minerals in, 44
Hay, 85
  chopped, 101
  evaluation, 99
  hay to grain ratios, 156
  how much?, 155
  sampling, 94
  time for harvesting, 100
Heaves, 215
Hemicellulose, 31
Hemoglobin, 220
Hoof growth, 215

Impaction, colon, 16, 22
Intestine, large, 17
Intestine, small, 15
Iodine, 58
  toxicity, 59
Iron, 52

Jejunum, 15

Kelp, 59

Lactation, 172
Lactose, 29
Lameness, *See* Founder
Laxative effects, 27
Lead, 64
Legumes, 85, 117
Lespediza, 108, 117
Licorice, 49
Lignin, 31
Limestone, 46
Lincomycin, 147
Linseed meal, 138
Lysine, 35

Magnesium, 47
Manganese, 59
Mares
  lactating, 172
  open, 164
  pregnant, 166
Mature horse, feeding, 158
Meat and bone meal, 140
Mercury, 65
Milk
  production, 172
  protein, 29, 140
  skim, 29
Milo, *See* Grain sorghum
Minerals, 39
  absorption, 16
  requirements, 164
  trace, 40
Molds, 100, 111, 117, 194
Molasses, 129
Molybdenum, 62
Monensin, 146
Moon blindness, 72
Mosquitoes, 221
Mouth, 12
Mules, 204

Nervousness, 158, 220
Niacin, 73
Nickel, 40
Nitrates, 109
Nitrites, 109
Nitrogen metabolism, 17
Nutrients
  interrelationship, 40
  requirements, 160
Nutritional secondary hyperpara-
    thyroidism, 10, 42

Oats
  feeding value, 123
  history, 1
Oat hay, 108
Ophthalmia, 72
Orchard grass, 118

Overfeeding, 186
Oxalate, 44, 104

Palatability, 130, 147
Pantothenic acid, 73
Paper as feed, 144
Parakeratosis, 61
Parasites, 21, 33
Pasture, 113
  management, 114
  species, 117
Peanut meal, 138
Peanut skins, 134
Pellets, 10, 143
Perosis, 59
Phosphorus, 41
  deficiency, 42
  soils, 46
  sources, 46
Phytic acid, 135
Phytobezoars, 23
Poisonous plants, 121
Potassium, 48
Potatoes, 132
Preferences, feed, 147
Protein
  deficiency, 35
  requirements, 37
  supplements, 6, 137
Prussic acid, 116
Pyridoxine, See Vitamin B6

Racehorses, 194
Rapeseed meal, 139
Ratio, Ca to P, 46, 104
Rations, examples for
  creep, 183
  weanling, 192
  mares, 173
  racehorses, 197
Recurrent uveitis, 72
Requirements
  energy, 158
  minerals, 158, 164
  protein, 158
  vitamins, 158, 164

Riboflavin, 72
Roaring, 71
Rumensin, 146
Rye, 128

Salt, 49
  iodized, 58
  trace mineralized, 51
Sand colic, 24
Selenium, 53
Silage, 111
Silicon, 40
Slobbering, moldy clover, 117
Snow, as source of water, 79
Sodium, 50
Sorghum, Grain, 127
Soybean meal, 137
Spelt, 129
Stallions, 176
Stomach, 14
Straws, 112
Sudan grass, 118
Sugar, preference, 150
Sulfur, 51
Sunflower meal, 139
Sweat, 4, 49, 103, 200
Sweet feed, 149
Synchronous diaphragmatic flutter, See Thumps

Teeth, floating, 12, 33
Tendons, contracted, 58, 59, 187
Theobromine, 133
Thiamin, 71, 220
Thumps, 200
Timothy
  feeding value, 109
  history, 5
  pasture value, 119
Tin, 40
Tonics, 7
Total digestible nutrients (TDN), 33
Trace minerals, 40
Trefoil, 106
Triticale, 129

Turfgrass clippings, 135, 139
Turnips, 131
Tying-up, 210

Ulcers, gastric, 15
Urea, 38
  utilization, 38
  toxicity, 39

Vitamins, 61
  A, 66
  B complex, 71
  B6, 74
  B12, 63, 72
  C, 74, 214
  D, 69
  E, 56, 70
  in feeds, 68
  K, 70, 107
  requirements, 158, 164
  supplements, 75
  synthesis in intestine, 18

Walnut shavings, 206
Water, 77
  absorption in, 72
  dehydration, 74, 200
  minerals in, 80
  requirements, 78
  temperature, 78
  toxic elements, 80
Weanlings, 182
Weight, birth, 178
Weight gains, 188
Weight tapes, 34
Wheat, 129
Wheat bran, 135
White muscle disease, 56
Wood chewing, 23, 145
Work requirements, 195

Xeropthalmia, 66

Yearlings, 194
Yeast, brewers, 72

Zinc, 61